ENERGY SCIENCE, ENGINEERING AND TECHNOLOGY

EXPLORING THE ENERGY-WATER NEXUS

Energy Science, Engineering and Technology

Additional books in this series can be found on Nova's website under the Series tab.

Additional E-books in this series can be found on Nova's website under the E-books tab.

Water Resource Planning, Development and Management

Additional books in this series can be found on Nova's website under the Series tab.

Additional E-books in this series can be found on Nova's website under the E-books tab.

ENERGY SCIENCE, ENGINEERING AND TECHNOLOGY

EXPLORING THE ENERGY-WATER NEXUS

PETER D. WRIGHT
EDITOR

Nova Science Publishers, Inc.
New York

Copyright ©2011 by Nova Science Publishers, Inc.

All rights reserved. No part of this book may be reproduced, stored in a retrieval system or transmitted in any form or by any means: electronic, electrostatic, magnetic, tape, mechanical photocopying, recording or otherwise without the written permission of the Publisher.

For permission to use material from this book please contact us:
Telephone 631-231-7269; Fax 631-231-8175
Web Site: http://www.novapublishers.com

NOTICE TO THE READER

The Publisher has taken reasonable care in the preparation of this book, but makes no expressed or implied warranty of any kind and assumes no responsibility for any errors or omissions. No liability is assumed for incidental or consequential damages in connection with or arising out of information contained in this book. The Publisher shall not be liable for any special, consequential, or exemplary damages resulting, in whole or in part, from the readers' use of, or reliance upon, this material. Any parts of this book based on government reports are so indicated and copyright is claimed for those parts to the extent applicable to compilations of such works.

Independent verification should be sought for any data, advice or recommendations contained in this book. In addition, no responsibility is assumed by the publisher for any injury and/or damage to persons or property arising from any methods, products, instructions, ideas or otherwise contained in this publication.

This publication is designed to provide accurate and authoritative information with regard to the subject matter covered herein. It is sold with the clear understanding that the Publisher is not engaged in rendering legal or any other professional services. If legal or any other expert assistance is required, the services of a competent person should be sought. FROM A DECLARATION OF PARTICIPANTS JOINTLY ADOPTED BY A COMMITTEE OF THE AMERICAN BAR ASSOCIATION AND A COMMITTEE OF PUBLISHERS.

Additional color graphics may be available in the e-book version of this book.

Library of Congress Cataloging-in-Publication Data

Exploring the energy-water nexus / editor: Peter D. Wright.
　　p. cm.
　Includes bibliographical references and index.
　ISBN 978-1-61209-791-6 (hardcover : alk. paper)
　1. Energy development--United States. 2. Electric power
production--United States. 3. Water consumption--United States. 4.
Oil-shale industry--United States--Water-supply. I. Wright, Peter D., 1960-
II. Title: Exploring the energy water nexus.
　TJ163.25.U6E97 2011
　331.91'23--dc22
　　　　　　　　　　　　　　　2011002528

Published by Nova Science Publishers, Inc. † New York

CONTENTS

Preface		vii
Chapter 1	Preliminary Observations on the Links between Water and Biofuels and Electricity Production *Anu Mittal*	1
Chapter 2	Improvements to Federal Water Use Data Would Increase Understanding of Trends in Power Plant Water Use *United States Government Accountability Office*	11
Chapter 3	Many Uncertainties Remain about National and Regional Effects of Increased Biofuel Production on Water Resources *United States Government Accountability Office*	63
Chapter 4	A Better and Coordinated Understanding of Water Resources Could Help Mitigate the Impacts of Potential Oil Shale Development *United States Government Accountability Office*	99
Chapter Sources		151
Index		153

PREFACE

Water and energy are inexorably linked - energy is needed to pump, treat and transport water, and large quantities of water are needed to support the development of energy. However, both water and energy may face serious constraints as demand for these vital resources continues to rise. This new book explores the energy-water nexus related to the national and regional effects of increased biofuel production on water resources; thermoelectric power plants and water and oil shale and water resources.

Chapter 1- Water and energy are inexorably linked—energy is needed to pump, treat, and transport water and large quantities of water are needed to support the development of energy. However, both water and energy may face serious constraints as demand for these vital resources continues to rise. Two examples that demonstrate the link between water and energy are the cultivation and conversion of feedstocks, such as corn, switchgrass, and algae, into biofuels; and the production of electricity by thermoelectric power plants, which rely on large quantities of water for cooling during electricity generation.

Chapter 2- In 2000, thermoelectric power plants accounted for 39 percent of total U.S. freshwater withdrawals. Traditionally, power plants have withdrawn water from rivers and other water sources to cool the steam used to produce electricity, so that it may be reused to produce more electricity. Some of this water is consumed, and some is discharged back to a water source.

In the context of growing demands for both water and electricity, this report discusses (1) approaches to reduce freshwater use by power plants and their drawbacks, (2) states' consideration of water use when reviewing proposals to build power plants, and (3) the usefulness of federal water data to experts and state regulators. GAO reviewed federal water data and studies on cooling technologies. GAO interviewed federal officials, as well as officials from seven selected states.

Chapter 3- In response to concerns about the nation's energy dependence on imported oil, climate change, and other issues, the federal government has encouraged the use of biofuels. Water plays a crucial role in all stages of biofuel production—from cultivation of feedstock through its conversion into biofuel. As demand for water from various sectors increases and places additional stress on already constrained supplies, the effects of expanded biofuel production may need to be considered.

Chapter 4- Oil shale deposits in Colorado, Utah, and Wyoming are estimated to contain up to 3 trillion barrels of oil—or an amount equal to the world's proven oil reserves. About

72 percent of this oil shale is located beneath federal lands, making the federal government a key player in its potential development. Extracting this oil is expected to require substantial amounts of water and could impact groundwater and surface water. GAO was asked to report on (1) what is known about the potential impacts of oil shale development on surface water and groundwater, (2) what is known about the amount of water that may be needed for commercial oil shale development, (3) the extent to which water will likely be available for commercial oil shale development and its source, and (4) federal research efforts to address impacts to water resources from commercial oil shale development. GAO examined environmental impacts and water needs studies and talked to Department of Energy (DOE), Department of the Interior (Interior), and industry officials.

Chapter 1

PRELIMINARY OBSERVATIONS ON THE LINKS BETWEEN WATER AND BIOFUELS AND ELECTRICITY PRODUCTION

Anu Mittal

WHY GAO DID THIS STUDY

Water and energy are inexorably linked—energy is needed to pump, treat, and transport water and large quantities of water are needed to support the development of energy. However, both water and energy may face serious constraints as demand for these vital resources continues to rise. Two examples that demonstrate the link between water and energy are the cultivation and conversion of feedstocks, such as corn, switchgrass, and algae, into biofuels; and the production of electricity by thermoelectric power plants, which rely on large quantities of water for cooling during electricity generation.

At the request of this committee, GAO has undertaken three ongoing studies focusing on the water-energy nexus related to (1) biofuels and water, (2) thermoelectric power plants and water, and (3) oil shale and water. For this testimony, GAO is providing key themes that have emerged from its work to date on the research and development and data needs with regard to the production of biofuels and electricity and their linkage with water. GAO's work on oil shale is in its preliminary stages and further information will be available on this aspect of the energy-water nexus later this year.

To conduct this work, GAO is reviewing laws, agency documents, and data and is interviewing federal, state, and industry experts. GAO is not making any recommendations at this time.

WHAT GAO FOUND

While the effects of producing corn-based ethanol on water supply and water quality are fairly well understood, less is known about the effects of the next generation of biofuel feedstocks. Corn cultivation for ethanol production can require from 7 to 321 gallons of water per gallon of ethanol produced, depending on where it is grown and how much irrigation is needed. Corn is also a relatively resource-intensive crop, requiring higher rates of fertilizer and pesticides than many other crops. In contrast, little is known about the effects of large-scale cultivation of next generation feedstocks, such as cellulosic crops. Since these feedstocks have not been grown commercially to date, there are little data on the cumulative water, nutrient, and pesticide needs of these crops and on the amount of these crops that could be harvested as a biofuel feedstock without compromising soil and water quality. Uncertainty also exists regarding the water supply impacts of converting cellulosic feedstocks into biofuels. While water usage in the corn-based ethanol conversion process has been declining and is currently estimated at 3 gallons of water per gallon of ethanol, the amount of water consumed in the conversion of cellulosic feedstocks is less defined and will depend on the process and on technological advancements that improve the efficiency with which water is used. Finally, additional research is needed on the storage and distribution of biofuels. For example, to overcome incompatibility issues between the ethanol and the current fueling and distribution infrastructure, research is needed on conversion technologies that can be used to produce renewable fuels capable of being used in the existing infrastructure.

With regard to power plants, GAO has found that key efforts to reduce use of freshwater at power plants are under way but may not be fully captured in existing federal data. In particular, advanced cooling technologies that use air, not water, for cooling the plant, can sharply reduce or even eliminate the use of freshwater, thereby reducing the costs associated with procuring water. However, plants using these technologies may cost more to build and witness lower net electricity output—especially in hot, dry conditions. Nevertheless, a number of power plant developers in the United States have adopted advanced cooling technologies, but current federal data collection efforts may not fully document this emerging trend. Similarly, plants can use alternative water supplies such as treated waste water from municipal sewage plants to sharply reduce their use of freshwater. Use of these alternative water sources can also lower the costs associated with obtaining and using freshwater when freshwater is expensive, but pose other challenges, including requiring special treatment to avoid adverse effects on cooling equipment. Alternative water sources play an increasingly important role in reducing power plant reliance on freshwater, but federal data collection efforts do not systematically collect data on the use of these water sources by power plants. To help improve the use of alternatives to freshwater, in 2008, the Department of Energy awarded about $9 million to examine among other things, improving the performance of advanced cooling technologies. Such research is needed to help identify cost effective alternatives to traditional cooling technologies.

Mr. Chairman and Members of the Subcommittee:

I am pleased to be here today to participate in your hearing on technology research and development for the energy-water linkage often referred to as the energy-water nexus. As you know, water and energy are inexorably linked, mutually dependant, and each affects the other's availability. Energy is needed to pump, treat, and transport water, and large quantities of water are needed to support the development of energy. Production of biofuels that may help reduce our dependency on oil, and the cooling of power plants that today provide the electricity we use, represent two examples where water supply is tied directly to our ability to provide energy.

However, both water and energy are facing serious supply constraints. Freshwater is increasingly in demand to meet the needs of municipalities, farmers, industries, and the environment. Likewise, rising demand for energy—fueled by both population growth and expanding uses of energy—may soon outstrip our ability to supply it with existing resources. Looking just at electricity, according to the Energy Information Administration's (EIA) most recent Annual Energy Outlook, 259 gigawatts of new generating capacity—the equivalent of 259 large coal-fired power plants—will be needed between 2007 and 2030. As the country's energy needs grow along with its population, additional pressure will likely be put on our water resources.

Given the importance of water and energy, both the federal government and state governments play key roles in monitoring, regulating, collecting information, and supporting research on energy and water issues. In general, state governments play a central role in overseeing water availability and use by evaluating water supplies and permitting water uses. However, while much of the authority governing water supply and distribution lies with state and local governments, the federal government also has a role in helping the country meet its energy needs without damaging or depleting our supplies of freshwater. For example, federal agencies, including the Department of Energy (DOE), have provided data and analysis about water use for energy production, as well as funded related research and development. These activities are important to further our understanding of how to more efficiently use such critical resources.

At the request of this committee, GAO currently has work under way related to three aspects of the energy-water nexus—water use in the production of biofuels, water use at thermoelectric power plants, and water use in the extraction of oil from shale. We expect to release reports on biofuels and thermoelectric power plants later this year. For each study, the committee asked us to identify technologies that could help reduce the amount of water needed to produce energy from these sources. My testimony today discusses key themes we have identified during our work to date on the two ongoing energy-water nexus jobs that are furthest along, specifically (1) biofuels and water use and (2) thermoelectric power plants and water use. Our work on oil shale is in its very preliminary stages and we will have further information to share with the committee on this aspect of the energy-water nexus later this year.

To identify the effects of biofuel cultivation, conversion, and storage on water supply and water quality, we are conducting a review of relevant scientific articles and key federal and state government reports. In addition, in consultation with the National Academy of Sciences, we identified and spoke with a number of experts who have published research analyzing the water supply requirements of one or more biofuel feedstocks and the implications of increased biofuel cultivation and conversion on water quality. Furthermore, we are interviewing officials from DOE, the Environmental Protection Agency (EPA), and the Department of Agriculture (USDA) about impacts on water supply and water quality during the cultivation of biofuel feedstocks and the conversion and storage of the finished biofuels.

To identify the relationship of thermoelectric plants and water, we are reviewing selected reports, interviewing federal officials and experts, and examining relevant energy and water data. In particular, we are examining reports on alternative cooling technologies and water supplies and the impact they can have on water use at power plants. We are also interviewing officials from DOE, EPA, and the Department of Interior's U.S. Geological Survey, as well as state water regulators and water and energy experts at national energy laboratories and universities. In addition, we are interviewing representatives from electric power producers, sellers of electric power plant equipment, cooling technology companies, and engineering firms that design new power plants. Finally, we are examining power plant data on water source, use, consumption, and cooling technology types collected by EIA and data collected and reported by the U.S. Geological Survey. Our work is being conducted in accordance with generally accepted government accounting standards. Those standards require that we plan and perform the audit to obtain sufficient, appropriate evidence to provide a reasonable basis for our findings and conclusions based on our audit objectives. We believe that the evidence obtained provides a reasonable basis for our findings and conclusions based on our audit objectives.

BACKGROUND

Biofuels are an alternative to petroleum-based transportation fuels and derived from renewable resources. Currently, most biofuels are derived from corn and soybeans. Ethanol is the most commonly produced biofuel in the United States, and about 98 percent of it is made from corn that is grown primarily in the Midwest. Corn is converted to ethanol at biorefineries through a fermentation process and requires water inputs and outputs at various stages of the production process—from growth of the feedstock to conversion into ethanol. While ethanol is primarily produced from corn grains, next generation biofuels, such as cellulosic ethanol and algae-based fuels, are being promoted for various reasons including their potential to boost the nation's energy independence and lessen environmental impacts, including on water. Cellulosic feedstocks include annual or perennial energy crops such as switchgrass, forage sorghum, and miscanthus; agricultural residues such as corn stover (the cobs, stalks, leaves, and husks of corn plants); and forest residues such as forest thinnings or chips from lumber mills. Some small biorefineries have begun to process cellulosic feedstocks on a pilot-scale basis; however, no commercial-scale facilities are currently operating in the United States.[1] In light of the federal renewable fuel standard's requirements for cellulosic ethanol starting in 2010,[2] DOE is providing $272 million to support the cost of constructing four small biorefineries that will process cellulosic feedstocks. In addition, in recent years, researchers have begun to explore the use of algae as a biofuel feedstock. Algae produce oil that can be extracted and refined into biodiesel and has a potential yield per acre that is estimated to be 10 to 20 times higher than the next closest quality feedstock. Algae can be cultivated in open ponds or in closed systems using large raceways of plastic bags containing water and algae.

Thermoelectric power plants use a fuel source—for example, coal, natural gas, nuclear material such as uranium, or the sun— to boil water to produce steam. The steam turns a turbine connected to a generator that produces electricity. Traditionally, water has been withdrawn from a river or other water source to cool the steam back into liquid so it may be reused to produce additional electricity. Most of the water used by a traditional thermoelectric power plant is for this cooling process, but water may also be needed for other purposes in the plant such as for pollution control equipment. In 2000, thermoelectric power plants accounted for 39 percent of total U.S. freshwater withdrawals.[3] EIA annually reports data on the water withdrawals, consumption and discharges of power plants of a certain size, as well as some information on water source and cooling technology type. These data are used by federal agencies and other researchers in estimating the overall power plant water use and determining how this use has and will continue to change.

INFORMATION IS LIMITED ON THE WATER SUPPLY AND WATER QUALITY IMPACTS OF THE NEXT GENERATION OF BIOFUELS

Our work to date indicates that while the water supply and water quality effects of producing corn-based ethanol are fairly well understood, less is known about the effects of the next generation of feedstocks and fuels. The cultivation of corn for ethanol production can require substantial quantities of water—from 7 to 321 gallons per gallon of ethanol produced—depending on where it is grown and how much irrigation water is used.[4] Furthermore, corn is a relatively resource-intensive crop, requiring higher rates of fertilizer and pesticide applications than many other crops; some experts believe that additional corn production for biofuels conversion will lead to an increase in fertilizer and sediment runoff and in the number of impaired streams and other water bodies. Some researchers and conservation officials have told us that the impact of corn-based ethanol on water supply and water quality could be mitigated through research into developing additional drought-tolerant and more nutrient-efficient crop varieties thereby decreasing the amount of water needed for irrigation and the amount of fertilizer that needs to be applied. Furthermore, experts also mentioned the need for additional data on current aquifer water supplies and research on the potential of biofuel cultivation to strain these water sources.

In contrast to corn-based ethanol, our work to date indicates that much less is known about the effects that large-scale cultivation of cellulosic feedstocks will have on water supplies and water quality. Since potential cellulosic feedstocks have not been grown commercially to date, there is little information on the cumulative water, nutrient, and pesticide needs of these crops, and it is not yet known what agricultural practices will actually be used to cultivate these feedstocks on a commercial scale. For example, while some experts assume that perennial feedstocks will be rainfed, other experts have pointed out that to achieve maximum yields for cellulosic crops, farmers may need to irrigate these crops. Furthermore, because water supplies vary regionally, additional research is needed to better understand geographical influences on feedstock production. For example, the additional withdrawals in states relying heavily on irrigation for agriculture, such as Nebraska, may place new demands on the Ogallala Aquifer, an already strained resource from which eight states draw water. In addition, if agricultural residues—such as corn stover—are to be used, this could negatively affect soil quality, increase the need for fertilizer, and lead to increased sediment runoff to waterways. Considerable uncertainty exists regarding the maximum amount of residue that can be removed for biofuels production while maintaining soil and water quality. USDA, DOE, and some academic researchers are attempting to develop new projections on how much residue can be removed without compromising soil quality, but sufficient data are not yet available to inform their efforts, and it may take several years to accumulate such data and disseminate it to farmers for implementation. Experts we spoke with generally agree that more research on how to produce cellulosic feedstocks in a sustainable way is needed.

Our work also indicates that even less is known about newer biofuels feedstocks such as algae. Algae have the added advantage of being able to use lower-quality water for cultivation, according to experts. However, the impact on water supply and water quality will ultimately depend on which cultivation methods are determined to be the most viable. Therefore, research is needed on how best to cultivate this feedstock in order to maximize its potential as a biofuel feedstock and limit its potential impacts on water resources. Other areas we have identified that relate to water and algae cultivation in need of additional research include:

- *Oil extraction*. Additional research is needed on how to extract the oil from the algal cell in such a way as to preserve the water contained in the cell along with the oil, thereby allowing some of that water to be recycled back into the cultivation process.
- *Contaminants*. Information is needed on how to manage the contaminants that are found in the algal cultivation water and how any resulting wastewater should be handled.

Uncertainty also exists regarding the water supply impacts of converting feedstocks into biofuels. Biorefineries require water for processing the fuel and need to draw from existing water resources. Water consumed in the corn-ethanol conversion process has declined over time with improved equipment and energy efficient design, according to a 2009 Argonne National Laboratory study, and is currently estimated at 3 gallons of water required for each gallon of ethanol produced. However, the primary source of freshwater for most existing corn ethanol plants is from local groundwater aquifers and some of these aquifers are not readily replenished. For the conversion of cellulosic feedstocks, the amount of water consumed is less defined and will depend on the process and on technological advancements that improve the efficiency with which water is used. Current estimates range from 1.9 to 5.9 gallons of water, depending on the technology used. Some experts we spoke with said that greater research is needed on how to manage the full water needs of biorefineries and reduce these needs further. Similar to current and next generation feedstock cultivation, additional research is also needed to better understand the impact of biorefinery withdrawals on aquifers and to consider potential resource strains when siting these facilities.

Our work to date also indicates that additional research is needed on the storage and distribution of biofuels. Ethanol is highly corrosive and poses a risk of damage to pipelines, and underground and above-ground storage tanks, which could in turn lead to releases to the environment that may contaminate groundwater, among other issues. These leaks can be the result of biofuel blends being stored in incompatible tank systems—those that have not been certified to handle fuel blends containing more than 10 percent ethanol. While EPA currently has some research under way, additional study is needed into the compatibility of higher fuel blends, such as those containing 15 percent ethanol, with the existing fueling infrastructure. To overcome potential compatibility issues, future research is needed on other conversion technologies that can be used to produce renewable and advanced fuels that are capable of being used in the existing infrastructure.

KEY EFFORTS TO REDUCE USE OF FRESHWATER AT POWER PLANTS MAY NOT BE FULLY CAPTURED IN EXISTING FEDERAL DATA

In our work to date, we have found (1) the use of advanced cooling technologies can reduce freshwater use at thermoelectric power plants, but federal data may not fully capture this industry change; (2) the use of alternative water sources can also reduce freshwater use, but federal data may not systematically capture this change; and (3) federal research under way is focused on examining efforts to reduce the use of freshwater in thermoelectric power plants.

Advanced cooling technologies offer the promise to reduce freshwater use by thermoelectric power plants. Unlike traditional cooling technologies that use water to cool the steam in power plants, advanced cooling technologies carry out all or part of the cooling process using air. According to power plant developers, they consider using these water-conserving technologies in new plants, particularly in areas with limited available water supplies. While these technologies can significantly reduce the amount of water used in a plant—and in some cases eliminate the use of water for cooling—their use entails a number of challenges. For example, plants using advanced cooling technologies may cost more to build and operate; require more land; and, because these technologies can consume a significant amount of energy themselves, witness lower net electricity output—especially in hot, dry conditions. However, eliminating or minimizing freshwater use by incorporating an advanced cooling technology provides a number of potential benefits to plant developers, including minimizing the costs associated with acquiring, transporting, and treating water, as well as eliminating impacts on the environment associated with water withdrawals, consumption, and discharge. In addition, the use of these advanced cooling technologies may provide the flexibility to build power plants in locations not near a source of water.

For these reasons, a number of power plant developers in the United States and across the world have adopted advanced cooling technologies, but according to EIA officials, the agency's forms have not been designed to collect information on the use of advanced cooling technologies. Moreover, the instruments the agency uses to collect these data were developed many years ago and have not been recently updated. EIA officials have told us that while some plants may choose to report this information, they may not do so consistently or in such a way that allows comprehensive identification of the universe of plants using advanced cooling technologies. Water experts and federal agencies we spoke to during the course of our work identified value in the annual EIA data on cooling technologies, but some explained that not having data on advanced cooling technologies limits public understanding of their prevalence and analysis of the extent to which their adoption results in a significant reduction in freshwater use. According to EIA officials, the agency is currently redesigning the instrument it uses to collect these data and expects to begin using the revised instrument in 2011. In addition, during the course of our work we noted that in 2002, EIA discontinued reporting water-related data for nuclear power plants, including water use and cooling technology. As we develop our final report, we will be looking at various suggestions that we can make to DOE to improve its data collection efforts.

Our work to date also indicates that the use of alternative water sources can substantially reduce or eliminate the need to use freshwater for power plant cooling at an individual plant. Alternative water sources that may be usable for power plant cooling include treated effluent from sewage treatment plants; groundwater that is unsuitable for drinking or irrigation because it is high in salts or other impurities; industrial water, such as water generated when extracting minerals like oil, gas, and coal; and others. Use of these alternative water sources can ease the development process where freshwater sources are in short supply and lower the costs associated with obtaining and using freshwater when freshwater is expensive. Because of these advantages, alternative water sources play an increasingly important role in reducing power plant reliance on freshwater, but can pose challenges, including requiring special treatment to avoid adverse effects on cooling equipment, requiring additional efforts to comply with relevant regulations, and limiting the potential locations of power plants to those nearby an alternative water source. These challenges are similar to those faced by power plants that use freshwater, but they may be exacerbated by the lower quality of alternative water sources.

Power plant developers we spoke with told us they routinely consider use of alternative water sources when developing their power plant proposals. Moreover, a 2007 report by Argonne National Laboratory indicates that the use of treated municipal wastewater at power plants has become more common, with 38 percent of power plants after 2000 using reclaimed water. EIA collects annual data from power plants on their water use and water source. However, according to EIA officials, while some plants report using an alternative water source, many may not be reporting such information since EIA's data collection form was not designed to collect data on these freshwater alternatives. One expert we spoke with told us that not having data on the use of alternative water sources at power plants limits public understanding of these trends and the extent to which these approaches are effective in reducing freshwater use. As we develop our final report, we plan to also develop suggestions for DOE that can improve this data gathering process.

Power plant developers may choose to reduce their use of freshwater for a number of reasons, such as when freshwater is unavailable or costly to obtain, to comply with regulatory requirements, or to address public concern. However, a developer's decision to deploy an advanced cooling technology or an alternative water source depends on an evaluation of the tradeoffs between the water savings and other benefits these alternatives offer and the cost involved. For example, where water is unavailable or prohibitively expensive, power plant developers may determine that despite the challenges, advanced cooling technologies or alternative water sources offer the best option for getting a potentially profitable plant built in a specific area.

While private developers make key decisions on what types of power plants to build and where to build them, and how to cool them based on their views of the costs and benefits of various alternatives, government research and development can be a tool to further the use of alternative cooling technologies and alternative water supplies. In this regard, the Department of Energy's National Energy Technology Laboratory (NETL) plays a central role in DOE's research and development effort. In recent years, NETL has funded research and development projects through its Innovations for Existing Plants program aimed at minimizing the challenges of deploying advanced cooling technologies and using alternative water sources at existing plants, among other things. In 2008, DOE awarded about $9 million to support research and development of projects that, among other things, could improve the performance of advanced cooling technologies, recover water used to reduce emissions of air pollutants at coal plants for reuse, and facilitate the use of alternative water sources such as polluted water for cooling. Such research endeavors, if successful, could alter the trade-off analysis power plant developers conduct in favor of nontraditional alternatives to cooling.

CONCLUDING OBSERVATIONS

Ensuring sufficient supplies of energy and water will be essential to meeting the demands of the 21st century. This task will be particularly difficult, given the interdependency between energy production and water supply and water quality and the strains that both these resources currently face. DOE, together with other federal agencies, has a key role to play in providing key information, helping to identify ways to improve the productivity of both energy and water, partnering with industry to develop technologies that can lower costs, and analyzing what progress has been made along the way. While we recognize that DOE currently has a number of ongoing research efforts to develop information and technologies that will address various aspects of the energy-water nexus, our work indicates that there are a number of areas to focus future research and development efforts. Investments in these areas will provide information to help ensure that we are balancing energy independence and security with effective management of our freshwater resources.

Mr. Chairman that concludes my prepared statement, I would be happy to respond to any questions that you or other Members of the Subcommittee might have.

End Notes

[1] For example, Range Fuels has operated a pilot biorefinery in Denver, Colo., since 2008 that has successfully converted pine and hardwoods into cellulosic ethanol. The company plans to optimize the technologies from this pilot plant at its cellulosic biorefinery, expected to begin commercial-scale production in 2010. This biorefinery, located in Soperton, Ga., is targeted to produce approximately 100 million gallons of ethanol and mixed alcohols from wood byproducts when it is at full scale.

[2] The Energy Independence and Security Act of 2007, Pub. L. No. 110-140 (2007).

[3] Water consumed by thermoelectric power plants accounts for a smaller percentage.

[4] Wu, M., M. Mintz, M. Wang, and S. Arora. *Consumptive Water Use in the Production of Ethanol and Petroleum Gasoline.* Center for Transportation Research, Energy Systems Division, Argonne National Laboratory, January 2009.

Chapter 2

IMPROVEMENTS TO FEDERAL WATER USE DATA WOULD INCREASE UNDERSTANDING OF TRENDS IN POWER PLANT WATER USE

United States Government Accountability Office

WHY GAO DID THIS STUDY

In 2000, thermoelectric power plants accounted for 39 percent of total U.S. freshwater withdrawals. Traditionally, power plants have withdrawn water from rivers and other water sources to cool the steam used to produce electricity, so that it may be reused to produce more electricity. Some of this water is consumed, and some is discharged back to a water source.

In the context of growing demands for both water and electricity, this report discusses (1) approaches to reduce freshwater use by power plants and their drawbacks, (2) states' consideration of water use when reviewing proposals to build power plants, and (3) the usefulness of federal water data to experts and state regulators. GAO reviewed federal water data and studies on cooling technologies. GAO interviewed federal officials, as well as officials from seven selected states.

WHAT GAO RECOMMENDS

To improve federal data collection efforts, GAO is making several recommendations, including that EIA consider collecting and reporting data on power plants' use of advanced cooling technologies and alternative water sources, and that USGS consider reinstating collection of data on power plant water consumption and distributing data on the use of alternative water sources. USGS agreed with our recommendations. DOE provided technical comments that we incorporated, as appropriate.

WHAT GAO FOUND

Advanced cooling technologies that rely on air to cool part or all of the steam used in generating electricity and alternative water sources such as treated effluent can reduce freshwater use by thermoelectric power plants. Use of such approaches may lead to environmental benefits from reduced freshwater use, as well as increase developer flexibility in locating a plant. However, these approaches also present certain drawbacks. For example, the use of advanced cooling technologies may result in energy production penalties and higher costs. Similarly, the use of alternative water sources may result in adverse effects on cooling equipment or regulatory compliance issues. Power plant developers must weigh these drawbacks with the benefits of reduced freshwater use when determining which approaches to pursue.

Consideration of water use by proposed power plants varies in the states GAO contacted, but the extent of state oversight is influenced by state water laws, related state regulatory policies, and additional layers of state regulatory review. For example, California and Arizona—states that historically faced constrained water supplies, have taken formal steps aimed at minimizing freshwater use at power plants. In contrast, officials in five other states GAO contacted said that their states had not developed official policies regarding water use by power plants and, in some cases, did not require a state permit for water use by new power plants.

Federal agencies collect national data on water availability and water use; however, of these data, state water agencies rely on federal water availability data when evaluating power plants' proposals to use freshwater more than federal water use data. Water availability data are collected by the U.S. Geological Survey (USGS) through stream flow gauges, groundwater studies, and monitoring stations. In contrast, federal data on water use are primarily used by experts, federal agencies, and others to identify industry trends. However, these data users identified limitations with the federal water use data that make them less useful for conducting trend analyses and tracking industry changes. For example, the Department of Energy's (DOE) Energy Information Administration (EIA) does not systematically collect information on the use of advanced cooling technologies and other data it collects are incomplete. Similarly, USGS discontinued distribution of data on water consumption by power plants and now only provides information on water withdrawals. Finally, neither EIA nor USGS collect data on power plant developers' use of alternative water sources, which some experts believe is a growing trend in the industry. Because federal data sources are a primary source of national data on water use by various sectors, data users told GAO that without improvements to these data, it becomes more difficult for them to conduct comprehensive analyses of industry trends and limits understanding of changes in the industry.

ABBREVIATIONS

CEC   California Energy Commission
DOE   Department of Energy
EIA   Energy Information Administration

EPA   Environmental Protection Agency
USGS   U.S. Geological Survey

October 16, 2009

The Honorable Bart Gordon
Chairman
Committee on Science and Technology
House of Representatives

Water and electricity are inexorably linked and mutually dependent, with each affecting the other's availability. Electricity is required to supply, purify, distribute, and treat water and wastewater; water is needed to generate electricity and to extract and process fuels used to generate electricity. Freshwater and electricity are important to our health, quality of life, and economic growth, and demand for both of these resources is rising. Freshwater is increasingly in demand to meet the needs of the public in growing cities and suburbs, farms, industries, and for recreation and wildlife. At the same time, electricity demand is projected to continue to grow in the United States, with the Department of Energy (DOE) estimating that U.S. electricity consumption will increase by an average of about 1 percent each year from 2007 through 2030. Limited availability of freshwater may make it more difficult to build new power plants, particularly in communities concerned about the adequacy of their water supply and maintaining the quality of aquatic environments. Periodic water shortages may also make it difficult for existing plants to satisfy demand for electricity. In recent years, water shortages and high water temperatures have caused reductions in electricity production at power plants in the United States and abroad, according to news reports.

In 2007, around three-fourths of the United States' electricity generating capacity consisted of thermoelectric power plants, which rely heavily on water for cooling. Thermoelectric power plants use a fuel source—for example, coal, natural gas, nuclear material such as uranium, or the sun—to boil water (boiler water) to produce steam. The steam turns a turbine connected to a generator that produces electricity. The steam is then cooled back into boiler water, a process which traditionally involves transferring heat from the steam to a separate water source (cooling water) and reusing it. Because the cooling water takes on the heat of the boiler water, some of it may evaporate, and the amount that evaporates varies, depending on the type of cooling technology that is used. In recent years, the majority of new thermoelectric power generating units have been combined cycle units, which use two processes to produce electricity, one of which is thermoelectric. In this type of plant, electricity is first generated by a simple cycle turbine that turns a generator directly as a result of burning fuel in the turbine—similar to jet engines used in aircraft. The heat produced by the simple cycle turbine that would otherwise be released to the atmosphere is used to produce steam which turns a steam turbine connected to a generator to produce electricity. Because some of the electricity is generated via the simple cycle turbine—a non-thermoelectric process—combined cycle plants use less water for cooling than similarly sized plants using only steam to produce electricity. Non-thermoelectric power plants, which accounted for the other one-quarter of 2007 U.S. electricity generating capacity, do not use water for cooling but still require water for other plant purposes, such as water for improving turbine performance on non-thermoelectric natural gas plants, as well as water for housekeeping activities.

Water use by thermoelectric power plants can be generally characterized as withdrawal, consumption, and discharge. Water withdrawals refer to water removed from the ground or diverted from a surface water source—for example, an ocean, river, or lake—for use by the plant. In 2000, the most recent USGS data available, thermoelectric power plants accounted for 39 percent of total U.S. freshwater withdrawals. Water consumption refers to the portion of the water withdrawn that is no longer available to be returned to a water source, such as when it has evaporated. In 1995, the most recent USGS data available, thermoelectric power plants accounted for 3 percent of freshwater consumption in the United States. Discharge refers to the return of water to its original source or a new source and represents the difference between withdrawals and consumption. For many thermoelectric power plants, much of the water they withdraw is later discharged, although often at higher temperatures. The amount of water discharged from a thermoelectric power plant depends on a number of factors, including the type of cooling technology used, plant economics, and environmental regulations.

Decisions to build a new power plant may be made independently by the power plant developer or with the consent of a state public utility commission. In either case, power plant developers must obtain approval from a number of state and local officials, generally by obtaining preconstruction and operating permits, before they can proceed with building their plant in a particular location. This process is meant to balance any adverse impacts a power plant may have on nearby communities and environments with the benefits it provides, such as energy supply and jobs. This regulation of the electricity industry's water use is complex and involves both state and federal laws. States are primarily responsible for managing the allocation and use of freshwater supplies. However, federal laws provide for control over the use of water in specific cases, such as on federal lands or in interstate commerce. In addition to the water power plants may withdraw, for which developers have to seek permits or purchase a water right, power plants may have to obtain permits to discharge water, since water discharged from a plant is regulated by the federal government and the states to ensure that it meets certain quality standards and does not harm protected species.[1] In some cases, plants may design their operations so they discharge no water into sources outside the plant boundaries, known as zero-liquid discharge.

Two federal agencies—the Department of the Interior's U.S. Geological Survey (USGS) and the Energy Information Administration (EIA), the independent statistical and analytical agency within DOE—collect key data that address how power plants use water. In addition, Congress recently passed the Omnibus Public Land Management Act of 2009, which included provisions known as the Secure Water Act.[2] The law authorizes, among other things, additional funding for the Department of the Interior to report water data to Congress, including thermoelectric power plant withdrawal data. Congress is also considering pending legislation related to energy and water. The Energy and Water Integration Act of 2009, among other things, calls for the National Academy of Sciences to conduct an analysis of the impact of energy development and production on U.S. water resources, including an assessment of water used in electricity production.[3] Similarly, the Energy and Water Research Integration Act directs DOE to take such steps as advancing energy and energy efficiency technologies that minimize freshwater use, increase water use efficiency, and utilize alternative water sources.[4] It also provides for the creation of a council to enhance energy and water resource data collection, including improving data on trends in power plant water use, among other things.

Because of the importance of freshwater to the public and society at large, the environment, and many industries, information about the country's current and expected use of freshwater and electricity is critical to making appropriate decisions about how these resources are managed. In this context, you asked us to provide information about the relationship between water and energy, which we will be addressing in several reports.[5] This report discusses water use in electricity production. More specifically, this report (1) describes technologies and other approaches to help reduce freshwater use by power plants and what, if any, drawbacks there are to using them, (2) describes the extent to which selected states consider water impacts of power plants when reviewing power plant development proposals, and (3) evaluates the usefulness of federal water data to experts and state regulators who evaluate power plant development proposals. We focused our evaluation on thermoelectric power plants, such as nuclear, coal, and certain natural gas plants. We did not consider the water supply issues associated with hydroelectric power, since the process through which hydroelectric plants use water is substantially different from that of thermoelectric plants and water is used to generate hydroelectric power without being directly consumed. We also limited our review to water used during the production of electricity at power plants and did not include water issues associated with extracting fuels used to produce electricity.

To understand technologies or other approaches to help reduce freshwater use by power plants and what, if any, drawbacks there are to using them, we reviewed industry, federal, and academic studies on alternative water sources and advanced cooling technologies that discussed these alternatives' benefits, as well as their drawbacks. We discussed the trade-offs associated with the use of these alternatives with power plant and cooling system manufacturers, U.S. national laboratory staff, academics, and other industry experts. To determine the extent to which selected states consider water impacts of power plants when reviewing power plant development proposals, we conducted case study reviews of three states: Arizona, California, and Georgia. We selected these states because of their differences in water availability and water law, high energy production, and large population centers. For each of these states, we met with state water regulators and siting authorities, power plant developers, water research institutions, and other subject matter experts. We also reviewed state water laws and policies for power plant water use. To supplement our case studies, we spoke with water regulators from four additional states: Nevada and Alabama, which shared watersheds with the case study states, and Illinois and Texas, which are large electricity producing states with sizable population centers. We did not attempt to determine whether states' efforts were reasonable or effective, rather, we only describe what states do to consider water impacts when making power plant siting decisions. To understand the usefulness of federal water data to experts and state regulators who evaluate power plant development proposals, we reviewed data and analysis from USGS and DOE's EIA and National Energy Technology Laboratory. We also conducted interviews about the usefulness of federal data with data users, including federal agencies; regulators from state departments of water resources and public utility commissions; and experts from environmental and water organizations, industry, and academia. A more detailed description of our scope and methodology is presented in appendix I.

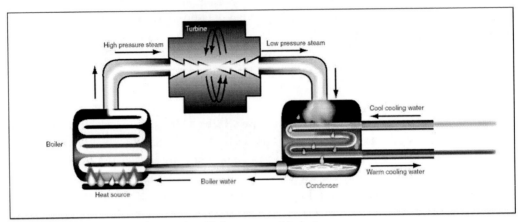

Source: GAO analysis of various national laboratory and industry sources.

Figure 1. Diagram of a Boiler Water Loop in a Power Plant.

We conducted this performance audit from October 2008 to October 2009, in accordance with generally accepted government auditing standards. Those standards require that we plan and perform the audit to obtain sufficient, appropriate evidence to provide a reasonable basis for our findings and conclusions based on our audit objectives. We believe that the evidence obtained provides a reasonable basis for our findings and conclusions based on our audit objectives.

BACKGROUND

Power plant developers consider many factors when determining where to locate a power plant, including the availability of fuel, water, and land; access to electrical transmission lines; electricity demand; and potential environmental issues. Often, developers will consider several sites that meet their minimum requirements, but narrow their selection based on economic considerations such as the cost of accessing fuel, water, or transmission lines, or the costs of addressing environmental factors at each specific site.

One key requirement for thermoelectric power plants is access to water. Thermoelectric power plants use a heat source to make steam, which is used to turn a turbine connected to a generator that makes electricity. As shown in figure 1, the water used to make steam (boiler water) circulates in a closed loop. This means the same water used to make steam is also converted back to liquid water —referred to as condensing—in a device called a condenser and, finally, moved back to the heat source to again make steam. In typical thermoelectric plants, water from a separate source, known as cooling water, flows through the condenser to cool and condense the steam in the closed loop after it has turned the turbine.

Consideration of water availability during the power plant siting process can pose different challenges in different parts of the country because precipitation and, relatedly, water availability varies substantially across the United States. Figure 2 shows the total amount of freshwater withdrawn in the United States as a percentage of available precipitation. Areas where the percentage is greater than 100—where more water is withdrawn than locally renewed through precipitation—are indicative of basins using other water sources transported by natural rivers and manmade flow structures, or may indicate unsustainable groundwater use.

Power plants can use various types of water for cooling—such as freshwater or saline water—and different water sources, including surface water, groundwater, and alternative water sources. An example of alternative water sources is reclaimed water such as treated effluent from sewage treatment plants. To make siting decisions, power plant developers typically consider the water sources that are available and least costly to use. Fresh surface water is the most common water source for power plants nationally, as shown in table 1.

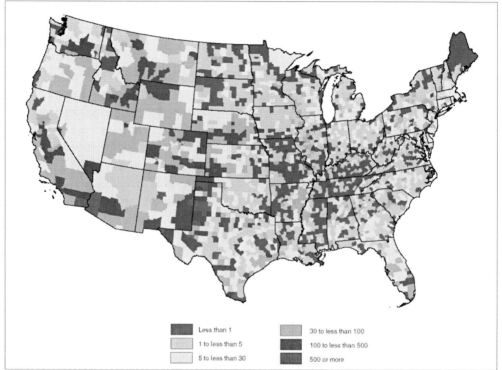

Source: Electric Power Research Institute. *A Survey of Water Use and Sustainability in the United States With a Focus on Power Generation.* (Palo Alto, CA. 2003.) 1005474; Map (Mapinfo).

Note: According to an Electric Power Research Institute official, the organization plans to update this analysis once USGS publishes 2005 freshwater withdrawal data.

Figure 2. Total Freshwater Withdrawal in 1995 as a Percentage of Available Precipitation.

Table 1. Estimated Water Withdrawals by Thermoelectric Power Plants in the United States in 2000

Millions of gallons per day	Surface Water	Groundwater
Saline water	59,500	0
Freshwater	135,000	409

Source: U.S. Geological Survey, *Estimated Use of Water in the United States in 2000*, (Reston, Virginia, 2004).

Cooling Technologies

Power plant developers must also consider what cooling technologies they plan to use in the plant. There are four general types of cooling technologies. Traditional cooling technologies that have been used for decades include once-through and wet recirculating cooling systems. Advanced cooling technologies that have focused on reducing the amount of cooling water used are relatively newer in the United States and include dry cooling and hybrid cooling. Specifically:

Once-through cooling systems. In once-through cooling systems, large amounts of cooling water are withdrawn from a water body such as a lake, river, or ocean, and used in the cooling loop. As shown in figure 3, the cooling water passes through the tubes of a condenser. As steam in the boiler water loop exits the turbine, it passes over the condenser tubes. This contact with the condenser tubes cools and condenses the steam back into boiler water for reuse. After the cooling water passes through the condenser tubes, it is discharged back into the water body warmer than it was when it was withdrawn.[6] Once-through cooling systems withdraw a significant amount of water but directly consume almost no water. However, because the water discharged back into the water body is warmer, experts believe that once-through systems may increase evaporation from the receiving water body. Furthermore, because of concerns about the harm withdrawal for once-through systems can have on aquatic life—when aquatic organisms are pulled into cooling systems, trapped against water intake screens, or their habitat is adversely affected by warm water discharges—these systems are rarely installed at new plants.

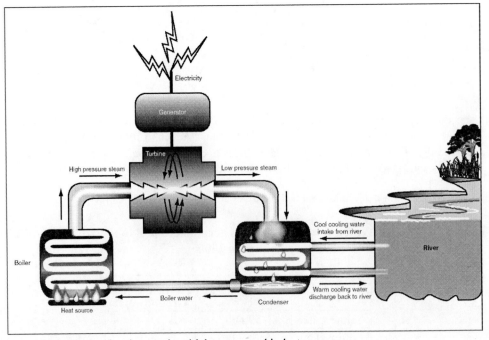

Source: GAO analysis of various national laboratory and industry sources.

Figure 3. Diagram of a Once-through Cooling System.

Improvements to Federal Water Use Data Would Increase Understanding of Trends... 19

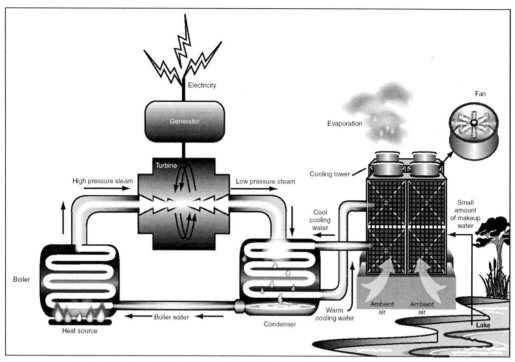

Source: GAO analysis of various national laboratory and industry sources.

Figure 4. Diagram of a Wet Recirculating System with a Cooling Tower.

Wet recirculating systems. Wet recirculating systems differ from once-through cooling systems in that they reuse cooling water multiple times. The most common type of recirculating system, shown in figure 4, uses cooling towers to dissipate the heat from the cooling water to the atmosphere. Similar to the once-through system, steam exiting the turbine is brought in contact with the tubes of a condenser that contain cooling water. The cooling water condenses the steam back into water for reuse in the boiler. The cooling water, warmed from the condenser, is then pumped to a cooling tower where it is exposed to the air. The heat from the warm cooling water is transferred to air flowing through the cooling tower, primarily through evaporation. In this process, some of the warm cooling water is consumed as it evaporates from the cooling tower, but most of it is returned to the condenser and used again. Over time, the quality of the cooling water is diminished as minerals and other dissolved and suspended solids present in the water are concentrated because of the water lost to evaporation. A portion of the cooling water containing the minerals and other dissolved solids must be discharged (known as blowdown) to prevent accumulation of those minerals and dissolved solids in the condenser, which could have adverse effects on condenser and cooling tower performance. For example, the National Energy Technology Laboratory estimated that a 520 megawatt wet recirculating system with a cooling tower circulates approximately 188,000 gallons of cooling water per minute. It withdraws around 5,000 gallons of water per minute to make up for the nearly 4,000 gallons per minute consumed through evaporation and approximately 1,000 gallons per minute discharged in the blowdown process. Some wet recirculating plants do not use a cooling tower but, instead, discharge cooling water to a pond, allowing it to cool before it is returned to the plant for reuse. For a wet recirculating system, water is only withdrawn from a water body to replace cooling water lost through evaporation and blowdown; thus, considerably less water is withdrawn than in a once-through cooling system. As a result, plants equipped with wet recirculating systems have relatively low water withdrawal but higher direct water consumption compared to once-through systems.

Dry cooling systems. Dry cooling systems rely primarily on air, rather than water, for cooling. In dry cooling systems, steam exiting the turbine flows through condenser tubes and is cooled directly by fans blowing air across the outside of these tubes to condense the steam back into liquid water. The cooled boiler water can then be reheated into steam to turn the turbine. In this approach, water is not used for cooling, although water still may be used for other plant purposes, such as pollution control equipment. As with the other systems, the steam, once cooled back into liquid water, is returned to the turbine for reuse.[7] See figure 5 for an illustration of dry cooling.

Hybrid cooling systems. Hybrid cooling technology offers a middle-ground option between wet and dry cooling systems, where wet and dry cooling components can be used either separately or simultaneously, as shown in figure 6. The system can operate both the wet and dry components in unison to increase cooling efficiency or may rely only on dry cooling to conserve water as needed.[8]

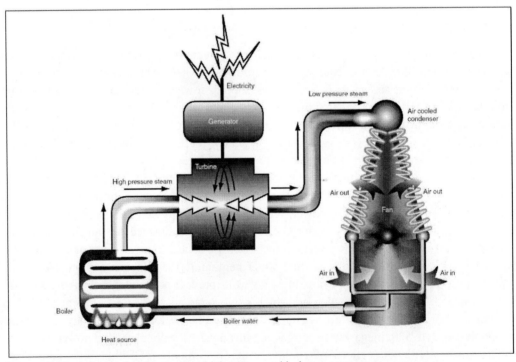

Source: GAO analysis of various national laboratory and industry sources.

Figure 5. Diagram of a Dry Cooling System.

Improvements to Federal Water Use Data Would Increase Understanding of Trends... 21

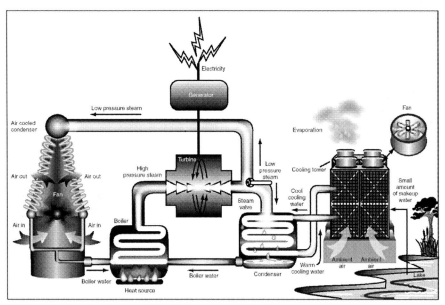

Source: GAO analysis of various national laboratory and industry sources.

Figure 6. Diagram of a Hybrid Cooling System.

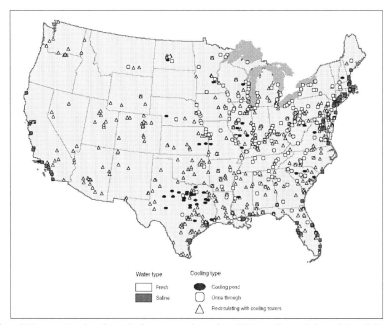

Source: National Energy Technology Laboratory, based on EIA-collected data; Map (Mapinfo).

Note: The National Energy Technology Laboratory developed this graphic based on 2000 and 2005 data collected by EIA and, as a result, power plants with a capacity less than 100 megawatts are not shown. According to an official from the National Energy Technology Laboratory, it was not possible using EIA data to determine the water type of cooling ponds. Additionally, as discussed later in the report, it is not possible to use EIA data to comprehensively identify the universe of plants with dry or hybrid cooling systems.

Figure 7. Water Based Cooling Systems by Technology and Water Source.

In 2008, the National Energy Technology Laboratory—a U.S. DOE laboratory that conducts and implements science and technology research and development programs in energy—estimated that 42.7 percent of U.S. thermoelectric generating capacity uses once-through cooling, 41.9 percent uses cooling towers, 14.5 percent uses cooling ponds, and 0.9 percent uses dry cooling.[9] Figure 7 illustrates the prevalence of different cooling technologies across the United States.

Federal Data Collection

Although a number of federal agencies collect data on water, two collect key data that are used to analyze the impacts of thermoelectric power plants and water availability: USGS and EIA.

- USGS's mission is to provide reliable scientific information to manage water, energy and other resources, among other things. USGS collects surface water and groundwater availability data through a national network of stream gauges and groundwater monitoring stations. USGS currently monitors surface and groundwater availability with approximately 7,500 streamflow gauges and 22,000 groundwater monitoring stations located throughout the United States.
- USGS compiles data and distributes a report every 5 years on national water use that describes how various sectors, such as irrigation, mining, and thermoelectric power plants, use water. USGS data related to thermoelectric power plants include (1) water withdrawal data at the state and county level organized by cooling technology—once-through and wet recirculating; (2) water source—surface or groundwater; and (3) whether water used was fresh or saline. USGS compiles water use data from multiple sources, including state water regulatory officials, power plant operators, and EIA. If data are not available for a particular state or use, USGS makes estimates.
- EIA's mission is to provide policy-neutral data, forecasts, and analyses to promote sound policy making, efficient markets, and public understanding regarding energy and its interaction with the economy and the environment. In carrying out this mission, EIA collects a variety of energy and electricity data nationwide, about topics such as energy supply and demand. For certain plants producing 100 megawatts or more of electricity, EIA collects data on water withdrawals, consumption, discharge, as well as some information on water source and cooling technology type. EIA annually collects water use data directly from power plants by using a survey.

State Water Laws

The variety of state water laws relating to the allocation and use of surface water can generally be traced to two basic doctrines the riparian doctrine, often used in the eastern United States, and the prior appropriation doctrine, often used in the western United States.

- Under the riparian doctrine, water rights are linked to land ownership—owners of land bordering a waterway have a right to use the water that flows past the land for any reasonable purpose. In general, water rights in riparian states may not be bought or sold. Landowners may, at any time, use water flowing past the land, even if they have never done so before. All landowners have an equal right to use the water, and no one gains a greater right through prior use. In some riparian states, water use is closely tracked by requiring users to apply for permits to withdraw water. In other states, where water has traditionally not been scarce, water use is not closely tracked. When there is a water shortage, water users share the shortage in proportion to their rights, or the amount they are permitted to withdraw, to the extent that it is possible to determine.
- Under the prior appropriation doctrine, water rights are not linked with land ownership. Instead, water rights are property rights that can be owned independent of land and are linked to priority and beneficial water use. A water right establishes a property right claim to a specific amount of water—called an allotment. Because water rights are not tied to land, water rights can be bought and sold without any ownership of land, although the rights to water may have specific geographic limitations. For example, a water right generally provides the ability to use water in a specific river basin taken from a specific area of the river. Water rights are also prioritized—water rights established first generally have seniority for the use of water over water rights established later—commonly described as "first in time, first in right." As a result, once established, water rights retain their priority for as long as they remain valid. For example, a water right to 100 acre feet of Colorado River water established in 1885 would retain that 1885 priority and allotment, even if the right was sold by the original party who established it. Water rights also must be exercised in order to remain valid, meaning rights holders must put the water to beneficial use or their right can be deemed abandoned and terminated—commonly referred to as "use it or lose it." When there is a water shortage in prior appropriation states, shortages fall on those who last obtained a legal right to use the water. As a result, a shortage can result in junior water rights holders losing all access to water, while senior rights holders have access to their entire allotment.

For some states, the legal framework for groundwater is similar to that of surface water as they use variants of either the riparian or prior appropriation doctrine to allocate water rights. However, in other states, the allocation of groundwater rights follows other legal doctrines, including the rule of capture doctrine and the doctrine of reasonable use. Under the rule of capture doctrine, landowners have the right to all the water they can capture under their land for any use, regardless of the effect on other water users. The doctrine of reasonable use similarly affords landowners the right to water underneath their land, provided the use is restricted to an amount necessary for reasonable use. In some cases, permits may be required prior to use and additional regulation may occur if a groundwater source is interconnected with surface water.

Power Plant Applications

A number of state agencies may be involved in considering or approving applications to build power plants or to use water in power plants. In some states, a centralized agency considers applications to build new power plants. In other states, applications may be filed with multiple state agencies. State water regulators issue water permits for power plants and other sectors to regulate water use and ensure compliance with relevant state laws and regulations. Public Utility Commissions, or the equivalent, may also have a role in authorizing the development of a power plant. In many states where retail electricity rates are regulated, these commissions are primarily responsible for approving the rates (or prices) electric utilities charge their customers and ensuring they are reasonable. As part of approving rates, these commissions approve utility investments into such things as new power plants and, as a result, may consider whether specific power plant design and cooling technologies are reasonable.

Thermoelectric Power Plants and Water Availability

Based on figures from EIA's 2009 Annual Energy Outlook, thermoelectric power plant generating capacity will increase by about 15 percent between 2006 and 2030. Depending on which cooling approaches are used, such an increase could further strain water resources. A variety of additional factors may also affect the availability of water for electricity generation and other uses, as well as the amount of water used to produce electricity. Some studies indicate that climate change will result in changes in local temperatures and more seasonal variations, both of which could cause increased levels of water consumption from thermoelectric power plant generation. Climate change may also result in changes in local precipitation and water availability, as well as more and longer droughts in some areas of the country. To the extent that this occurs, power plant operators may need to reduce the use of water for power plant cooling. In addition, some technologies aimed at reducing greenhouse gas emissions, such as carbon capture technologies, may require additional water. The combination of environmental laws, climate change, and the inclusion of new water intensive air emission technologies may impact water availability and require power plants operators to reduce water use in the future. In addition, since the water inlet structures used at once-through cooling plants can either trap or draw in fish and other aquatic life—referred to as impingement and entrainment—there is increased pressure to reduce the use of once-through cooling at existing plants.

ADVANCED COOLING TECHNOLOGIES AND ALTERNATIVE WATER SOURCES CAN REDUCE THE USE OF FRESHWATER AT POWER PLANTS, BUT THEIR ADOPTION POSES CERTAIN DRAWBACKS

Advanced cooling technologies and alternative water sources can reduce freshwater use by thermoelectric power plants, leading to a number of benefits for plant developers; however, incorporating each of these options for reducing freshwater use into thermoelectric power plants also poses certain drawbacks. Benefits of reducing freshwater use may include social and environmental benefits, minimizing water-related costs, as well as increasing a developer's flexibility in determining where to locate a new plant. On the other hand, drawbacks to using advanced cooling technologies may include potentially lower net electricity output, higher costs, and other trade-offs. Similarly, the use of alternative water sources, such as treated effluent or groundwater unsuitable for drinking or irrigation, may have adverse effects on cooling equipment, pose regulatory challenges, or be located too far from a proposed plant location to be a viable option. Power plant developers must weigh the trade-offs of these drawbacks with the benefits of reduced freshwater use when determining what approaches to pursue, and must consider both the economic costs over a plant's lifetime and the regulatory climate. For example, in a water-scarce region of the country where water costs are high and there is significant regulatory scrutiny of water use, a power plant developer may opt for a water-saving technology despite its drawbacks.

Advanced Cooling Technologies and Alternative Water Sources Can Reduce Freshwater Use, Leading to a Number of Benefits

Advanced cooling technologies under development and in limited commercial use and alternative water sources can reduce the amount of freshwater needed by plants, resulting in a number of benefits to both the environment and plant developers. As shown in table 2, dry cooling can eliminate nearly all the water withdrawn and consumed for power plant cooling.

Hybrid cooling systems, depending on design, can reduce water use—generally to a level between that of a wet recirculating system with cooling towers and a dry cooling system. According to the Electric Power Research Institute, hybrid systems are typically designed to use 20-80 percent of the water used for a wet recirculating system with cooling towers.[10]

In addition to using advanced cooling technologies, power plant operators can reduce freshwater use by utilizing water sources other than freshwater. Alternative water sources include treated effluent from sewage treatment plants; groundwater that is unsuitable for drinking or irrigation because it is high in salts or other impurities; sea water; industrial water and water generated when extracting minerals like oil, gas, and coal. For example, the oil and gas production process can generate wastewater, which is the subject of research as a possible source of cooling water for power plants.

Use of alternative water sources by power plants is increasing in some areas, and two power plant developers we spoke with said they routinely consider alternative water sources when planning new power plants, particularly in areas where water has become scarce, tightly regulated, or both. A 2007 report by the DOE's Argonne National Laboratory identified at least 50 power plants in the United States that use reclaimed water for cooling and other purposes, with Florida and California having the largest number of plants using reclaimed water.[11] According to the report, the use of reclaimed water at power plants has become more common, with 38 percent of the plants using reclaimed water doing so after 2000. One example of a power plant using an alternative to freshwater is Palo Verde, located near Phoenix, Arizona—the largest U.S. nuclear power plant, with a capacity of around 4,000 megawatts. Palo Verde uses approximately 20 billion gallons of treated effluent annually from treatment plants that serve several area municipalities, comprising over 1.5 million people.

Table 2. Selected Estimates of Water Withdrawn and Consumed for Power Plant Cooling by Cooling Technology and Plant Type[a]

Gallons per megawatt hour by type of plant	Once-through Withdrawal	Once-through Consumption[b]	Wet recirculating with cooling tower Withdrawal	Wet recirculating with cooling tower Consumption	Dry cooling Withdrawal	Dry cooling Consumption
Coal	20,000 – 50,000	300	500-600	480	0	0
Combined cycle	7,500-20,000	100	230	180	0	0
Nuclear	25,000 – 60,000	400	800-1,100	720	[c]	[c]
Solar thermal (trough)	—	—	600-850[d]	[d]	0	0

Sources: Coal, natural gas and nuclear estimates: Electric Power Research Institute, *Water and Sustainability (Volume 3): U.S. Water Consumption for Power Production—The Next Half Century*. (Palo Alto, CA, 2002). 1006786. Dry cooling and solar thermal: Electric Power Research Institute, *Water Use for Electric Power Generation*, (Palo Alto, CA, 2008). 1014026.

Note: We did not include water use estimates for hybrid cooling in this table, because these systems' water use is very dependent on their design and operation, including the proportion of wet versus dry cooling. Additionally, for wet recirculating systems, we provided water use estimates only for those systems with cooling towers, since according to work conducted by the National Energy Technology Laboratory, they are more common than wet recirculating systems with cooling ponds.

[a] In addition to cooling water, water may be used for other plant purposes, such as environmental controls; make-up boiler water; and water for cleaning, drinking, and sanitation. As a result, while dry and hybrid systems may eliminate or minimize water needs for cooling, total plant water use will not be eliminated entirely. Furthermore, some plants, such as natural gas simple cycle, solar photovoltaic, and wind, are not considered thermoelectric and do not use water for cooling but may use water for other plant purposes.

[b] Once-through cooling systems discharge water at a warm temperature; therefore, water consumption in these systems occurs via evaporation downstream of the plant.

[c] Representatives from one engineering firm and some power plant developers we spoke to explained that the large size of dry cooling systems needed for plants that derive all of their electricity production from the steam cycle, for example, nuclear and coal plants, may introduce challenges. Furthermore, according to another expert, one type of dry cooled technology may not be approved for use with certain nuclear reactors because of safety concerns.

[d] This estimate for solar thermal (trough) water withdrawals is from the Electric Power Research Institute's 2008 report. This report did not identify a comparable range for water consumption. Other sources we reviewed estimated water consumption rates for solar trough plants ranging from 740 gallons to 920 gallons per megawatt hour.

Reducing the amount of freshwater needed for cooling leads to a number of social and environmental benefits and may benefit developers by lowering water-related costs and providing more flexibility in choosing a location for a new plant, among other things.

Social and Environmental Benefits

Reducing the amount of freshwater used by power plants through the use of advanced cooling technologies and alternative water sources has the potential to produce a number of social and environmental benefits. For example, limiting freshwater use may reduce the impact to the environment associated with withdrawals, consumption, and discharge. Freshwater is in high demand across the United States. Reducing freshwater withdrawals and consumption by the electricity sector makes this limited resource more available for additional electricity production or competing uses, such as public water supplies or wildlife habitat. Furthermore, eliminating water use for cooling entirely, such as by using dry cooling, could minimize or eliminate the water discharges from power plants, a possible source of heat and pollutants to receiving water bodies, although regulations limit the amount of heat and certain pollutants that may be discharged into water bodies.

Water-Related Cost Savings

By eliminating or minimizing the use of freshwater for cooling, power plant developers may reduce some water-related costs, including the costs associated with acquiring, transporting, treating, and disposing of water. Depending on state water laws, a number of costs may be associated with acquiring water—purchasing a right to use water, buying land with a water source on or underneath it, or buying a quantity of freshwater from a municipal or other source. Eliminating the need to purchase water for cooling by using dry cooling could reduce these water-related expenses. Using an alternative water source, if less expensive than freshwater, could reduce the costs of acquiring water, although treatment costs may be higher. Power plant developers and an expert from a national laboratory told us the costs of acquiring an alternative water source are sometimes less than freshwater, but vary widely depending on its quality and location. In addition to lowering the costs associated with acquiring water, if water use for cooling is eliminated entirely, plant developers may eliminate the need for a pipeline to transport the water, as well as minimize costs associated with treating the water. Water-related costs are one of several costs that power plant developers will consider when evaluating alternatives to freshwater. Since the cost of freshwater may rise as demand for freshwater increases, a developer's ability to minimize power plant freshwater use could become increasingly valuable over time.

Siting Flexibility and Other Benefits

Minimizing or eliminating the use of freshwater may offer a plant developer increased flexibility in determining where to locate a power plant. According to power plant developers we spoke with, siting a power plant involves balancing factors such as access to fuel, including natural gas pipelines, and access to large transmission lines that carry the electricity produced to areas of customer demand. Some explained that finding a site that meets these factors and also has access to freshwater can be challenging. Power plant developers we spoke with said options such as dry cooling and alternative water sources have offered their companies the flexibility to choose sites without freshwater, but with good access to fuel and transmission.

According to power plant developers and an expert from a national laboratory we spoke with, eliminating or lowering freshwater use can lead to other benefits, such as minimizing regulatory hurdles like the need to acquire certain water permits. Furthermore, using a nonfreshwater source may be advantageous in areas with more regulatory scrutiny of or public opposition to freshwater use.

Adoption of Advanced Cooling Technologies May Reduce Electricity Production, Increase Costs, and Pose Other Drawbacks

Despite the benefits associated with the lower freshwater requirements of advanced cooling technologies, these technologies have a number of drawbacks related to electricity production and costs that power plant developers will have to consider during their decisionmaking process.

Energy Production Penalties

Despite the many benefits advanced cooling technologies offer, both dry cooling and hybrid cooling technologies may reduce a plant's net energy production to a greater extent than traditional cooling systems—referred to as an "energy penalty." Energy penalties result in less electricity available outside the plant, which can affect plant revenues, and making up for the loss of this electricity by generating it elsewhere can result in increases in water use, fuel consumption, and air emissions. Energy penalties result from (1) energy consumed to run cooling system equipment, such as fans and pumps, and (2) lower plant operating efficiency—measured as electricity production per unit of fuel—in hot weather due to lower cooling system performance. Specifically, energy penalties include:

- *Energy needed for cooling system equipment.* Cooling systems, like many systems in a power plant, use electricity produced at the plant to operate, which results in less electricity available for sale. According to experts we spoke with, because dry cooling systems and hybrid cooling systems rely on air flowing through a condenser, energy is needed to run fans that provide air flow, and the amount of energy needed to run cooling equipment will depend on such factors as system design, season, and region.[12] A 2001 EPA study estimated that for a combined cycle plant, energy requirements to operate a once-through system (pumps) are 0.15 percent of plant output, 0.39 percent of plant output for a wet recirculating system with cooling towers (pumps and fans), and 0.81 percent of plant output for a dry cooled system (fans).[13]
- *Plant operating efficiency and cooling system performance.* Plants using a dry cooling component, whether entirely dry cooled or in a hybrid cooled configuration, may face reduced operating efficiency under certain conditions. A power plant's operating efficiency is affected by the performance of the cooling system, among other things, and power plants with systems that cool more effectively produce electricity more efficiently. A cooling system's effectiveness is influenced both by

the design of the cooling system and ambient conditions that determine the temperature of that system's cooling medium—water in once-through and wet recirculating systems and air in dry cooling systems. In general, the effectiveness of a cooling system decreases as the temperature of the cooling medium increases, since a warmer medium can absorb less heat from the steam. Once-through systems cool steam using water being withdrawn from the river, lake, or ocean. Wet recirculating systems with cooling towers, on the other hand, use the process of evaporation to cool the steam to a temperature that approaches the "wet-bulb temperature"—an alternate measure of temperature that incorporates both the ambient air temperature and relative humidity. In contrast, dry cooled systems transfer heat only to the ambient air, without evaporation. As a result, dry cooled systems can cool steam only to a temperature that approaches the "dry-bulb temperature"—the measure of ambient air temperature measured by a standard thermometer and with which most people are familiar. In general, once-through systems tend to cool most effectively because the temperature of the body of water from which cooling water is drawn is, on average, lower than the wet- or dry-bulb temperature. Moreover, wet-bulb temperatures are generally lower than dry-bulb temperatures, often making recirculating systems more effective at cooling than dry cooled systems. Further, according to one report that we reviewed, greater fluctuations in dry-bulb temperatures seasonally and throughout the day can make dry cooled systems harder to design.[14] Dry bulb temperatures can be especially high in hot, dry parts of the country, such as the Southwest, leading to significant plant efficiency losses during periods of high temperatures, particularly during the summer. According to experts and power plant developers we spoke with, plant efficiencies may witness smaller reductions during other parts of the year when temperatures are lower or in cooler climates.[15] Nevertheless, in practice, lower cooling system performance can result in reduced plant net electricity output or greater fuel use if more fuel is burned to produce electricity to offset efficiency losses. Plant developers can take steps to reduce efficiency losses such as by installing a larger dry cooling system with additional cooling capability, but such a system will result in higher capital costs.

A plant's total energy penalty will be a combination of both effects described—energy needed for cooling system equipment and the impact of cooling system performance on plant operating efficiency. Energy penalties may result in lost revenue for the plant due to the net loss in electricity produced for a given unit of fuel, especially during the summer when electricity demand and prices are often the highest. Energy penalties may also affect the price consumers pay for electricity in a regulated market, if the cost of the additional fuel needed to produce lost electricity is passed on to consumers by regulators. Finally, energy penalties may affect emissions of pollutants and carbon dioxide if lost output is made up for by an emissions producing power plant, such as a coal- or natural gas-fueled power plant. This is because additional fuel is burned to produce electricity that offsets what was lost as a result of the energy penalty, and, thus, additional carbon dioxide and other pollutants are released.

Recent studies comparing total energy penalties between cooling systems have used differing methodologies to estimate energy penalties and have reached varying conclusions.[16] For example, a 2001 EPA study estimates the national average, mean annual energy penalties—lower electricity output—for plants operating at two-thirds capacity with dry cooling to be larger than those with wet recirculating systems with cooling towers. In this study, EPA estimated penalties of 1.7 percent lower output for a combined cycle plant with a dry system compared to a wet recirculating system with a cooling tower, and 6.9 percent lower output for a fossil fueled plant run fully on steam, such as a coal plant.[17] Similarly, a separate study conducted by two DOE national labs in 2002 estimated larger annual energy penalties for hypothetical 400 megawatt coal plants in multiple regions of the country retrofitted to dry cooling—these penalties ranged between 3 to 7 percent lower output on average for a plant retrofitted with a dry cooled system compared to a plant retrofitted with a wet recirculating system with a cooling tower. On the hottest 1 percent of temperature conditions during the year, this energy penalty rose to between 6 and 10 percent lower output for plants retrofitted to dry cooling compared with those retrofitted to a wet recirculating system with cooling towers.[18] However, some experts we spoke with told us energy penalties are higher in retrofitted plants than when a dry cooled system is designed according to the unique specifications of a newly built plant.

A 2006 study conducted for the California Energy Commission estimated electricity output and other characteristics for new, theoretical combined cycle natural gas plants in four climatic zones of California using different cooling systems. The study found that dry cooling systems result in significant water savings, but that plants using wet cooling systems generally experience higher annual net electricity output, as shown in table 3, and lower fuel consumption. Furthermore, while the study estimates that plant capacity to produce electricity is limited on hot days for both types of cooling systems, the hot day capacity of the dry cooled plant to produce electricity is up to 6 percent lower than the wet recirculating plant with cooling tower.[19]

Power plant developers can take steps to address the energy penalties associated with dry cooling technology by designing their plants with larger dry cooled systems capable of performing better during periods of high ambient temperatures. Alternatively, they can use a hybrid technology that supplements the dry system with a wet recirculating system with a cooling tower during the hottest times of the year. However, in making this decision, developers must weigh the trade-offs between the costs associated with building and operating a larger dry cooled system or a hybrid system and the benefits of lowering their energy penalties.

Higher Costs

According to some power plant developers and experts we spoke with, another drawback to using dry and hybrid cooling technologies is that these technologies typically have higher capital costs. Experts, power plant developers, and studies indicated that while capital costs for each system can vary significantly, as a general rule, capital costs are lowest for once-through systems, higher for wet recirculating systems, and highest for dry cooling. Some told us the capital costs of hybrid systems—as a combination of wet recirculating and dry cooling systems—generally fall in between these two systems. Furthermore, according to some of the experts we spoke with and studies we reviewed, the capital costs of a plant's cooling system vary based on the specific characteristics of a given plant, such as the costs of the cooling towers, the circulating water lines to transport water to and around the plant, pumps, fans, as well as the extent to which a dry cooled system is sized larger to offset energy penalties. As with energy penalties, studies estimating capital costs for dry and hybrid systems have used differing methodologies and provide varying estimates of capital costs.[20] One study by the

Electric Power Research Institute estimated dry cooling system capital costs for theoretical 500 megawatt combined cycle plants in 5 climatic locations to be 3.6 to 4.0 times that of wet recirculating systems with cooling towers.[21] Experts from an engineering firm we spoke with also explained that capital costs for dry and hybrid cooled systems can be many times that of a wet recirculating system with cooling towers. They estimated that, in general, installing a dry system on a 500 megawatt combined cycle plant instead of a wet recirculating system with a cooling tower could increase baseline capital costs by $9 to $24 million, depending on location—an increase in baseline capital costs that is 2.0 to 5.1 times higher than if a wet recirculating system with a cooling tower were used. They estimated dry cooling to be more costly on a 500 megawatt coal plant, with dry cooling resulting in an increase in baseline capital costs that was 2.6 to 7.0 times higher than if a wet recirculating system with a cooling tower were used.

With respect to annual costs, according to experts we spoke with and studies we reviewed, annual cost differences between alternative cooling technologies and traditional cooling technologies are variable and may depend on such factors as the costliness of obtaining and treating water, the extent to which cooling water is reused within the system, the need for maintenance, the extent to which energy penalties result in lost revenue, and the extent to which a cooling system is sized larger to offset energy penalties. Estimates from four reports we reviewed calculated varying cooling system annual costs for a range of plant types and locations using different methodologies, and found annual costs of dry systems to generally range from one and a half to four times those of wet recirculating systems with cooling towers. One of these studies, however, in examining the potential for higher water costs, found that dry cooling could be more economical on an annual basis in some areas of the country with expensive water or become more economical in the future if water costs were to rise.[22] Furthermore, an expert from an engineering firm we spoke with explained that cooling system costs are only one component of total plant costs, and that while one cooling system may be expensive relative to another, its impact on total plant costs may not be as significant in a relative sense if the plant's total costs are high.

Space, Noise, and Suitability Issues

There may be other drawbacks to dry cooled technology, including space and noise considerations. Towers, pumps, and piping for both dry cooled and wet cooled systems with cooling towers require substantial space, but according to experts we spoke with, dry cooled systems tend to be larger. For example, according to one expert we spoke with, a dry cooled system for a natural gas combined cycle plant that derives one-third of its electricity from the steam cycle could be almost as large as two football fields. Moreover, according to others, the large size of dry cooling systems needed for plants that derive all of their electricity production from the steam cycle—for example, nuclear and coal plants—may make the use of dry cooling systems less suitable for these kinds of power plants. Experts we spoke with explained that because full steam plants produce all of their electricity by heating water to make steam, they require larger cooling systems to condense the steam back into usable liquid water. As a result, the size of a dry cooling system for a full steam plant could be three times that of a dry cooling system for a similarly-sized combined cycle plant that only produces one-third of its electricity from the steam cycle.

Furthermore, according to one expert we spoke with, the most efficient type of dry cooled technology may not be approved for use with certain nuclear reactors, because of safety concerns. Finally, the motors, fans, and water of both dry cooled and wet recirculating systems with cooling towers may create noise that disturbs plant employees, nearby residents, and wildlife. Noise-reduction systems may be used to address this concern, although they introduce another cost trade-off that plant developers must consider.

Use of Alternative Water Sources May Also Pose Certain Drawbacks

Despite the growth in plants using alternative water sources, there are a number of drawbacks to using this water source instead of freshwater. While some of these drawbacks are similar to those faced by power plants that use freshwater, they may be exacerbated by the lower quality of alternative water sources. These drawbacks include adverse effects to cooling equipment, regulatory compliance issues, and access to alternative water sources, as follows.

Adverse Effects to Cooling Equipment

Water used in power plants must meet certain quality standards in order to avoid adverse effects to cooling equipment, such as corrosion, scaling, and the accumulation of micro or macrobiological organisms. While freshwater can also cause adverse effects, the generally lower quality of alternative water sources make them more likely to result in these effects. For example, effluent from a sewage treatment plant may be higher in ammonia than freshwater, which can cause damage to copper alloys and other metals. High levels of ammonia and phosphates can also lead to excessive biological growth on certain cooling tower structures. Chemical treatment is used to mitigate such adverse effects of alternative water sources when they occur, but this treatment results in additional costs. According to one power plant operator we spoke with, alternative water sources often require more extensive and expensive treatment than freshwater sources, and it can be a challenging process to determine the precise makeup of chemicals needed to minimize the adverse effects.

Regulatory Compliance Issues

Power plant developers using alternative water sources may face additional regulatory challenges. Depending on their design, power plants may discharge water directly to a water source, such as a surface water body, or release water into the air through cooling towers. As a result, power plants must comply with a number of water quality and air regulations, and the presence of certain pollutants in alternative water sources can make compliance more challenging. For example, reclaimed water from sewage treatment plants is treated to eliminate bacteria and other contaminants that can be harmful to humans. Similarly, water associated with minerals extraction may contain higher total dissolved and suspended solids and other constituents, which could adversely affect the environment if discharged. Addressing these issues through the following actions entail additional costs to the power plant operators: (1) chemical treatment prior to discharging water to another water source, (2) discharging water to a holding pond unconnected to another water source for evaporation, or (3) eliminating all liquid discharges by, for example, evaporating all the water used at the plant and disposing of the resulting solid waste into a facility such as a landfill.

Table 3. Percentage Difference in Annual Net Plant Electricity Output for Theoretical Combined Cycle Plants with Different Cooling Systems at Four Geographic Locations in California

Geographic locations	Percentage difference in annual net plant electricity output for a wet recirculating system with cooling towers compared to a dry cooled system
Desert (hot, arid)	1.07
Valley (hot, humid)	1.46
Coast (cool, humid)	0.37
Mountain (variable, elevated)	1.87

Source: Maulbetsch, J. S. and M.N. DiFilippo, *Cost and Value of Water Use at Combined-Cycle Power Plants*, California Energy Commission, PIER Energy-Related Environmental Research, CEC-500-2006-034. April 2006.

Access to Alternative Water Sources

As with freshwater sources, the proximity of an alternative water source may be a drawback that power plant developers have to consider when pursuing this option. Power plant developers wishing to use an alternative water source must either build the plant near that source—which can be challenging if that water source is not also near fuel and transmission lines—or pay the costs of transporting the water to the power plant's location, such as through a pipeline. Furthermore, two power plant developers we spoke with told us that certain alternative water sources, like treated effluent, are in increasing demand in some parts of the country, making it more challenging or costly to obtain than in the past.

Power Plant Developers Must Weigh Trade-offs When Evaluating Options to Reduce Freshwater Use

A power plant developer may want to reduce the use of freshwater for a number of reasons, such as when freshwater is unavailable or costly to obtain, to comply with regulatory requirements, or to address public concern. However, power plant developers we spoke with told us that when considering the viability of an advanced cooling technology or alternative water source, they must weigh the trade-offs between the water savings and other benefits these alternatives offer with the drawbacks to their use. For example, in a water-scarce region of the country where water costs are high and there is much regulatory scrutiny of water use, a power plant developer may determine that, despite the drawbacks associated with the use of advanced cooling technologies or alternative water sources, these alternatives still offer the best option for getting a potentially profitable plant built in a specific area. Furthermore, according to power plant developers we spoke with, these decisions have to be made on a project by project basis because the magnitude of benefits and drawbacks will vary depending on a plant's type, location, and the related climate. For example, dry cooling has been installed in regions of the country where water is relatively plentiful, such as the Northeast, to help shorten regulatory approval times and avoid concerns about the adverse impacts that other cooling technologies might have on aquatic life. In making a determination about what cooling technology to use, power plant developers evaluate the net economic costs of alternatives like dry cooling or an alternative water source—its savings compared to its costs—over the life of a proposed plant, as well as the regulatory climate. Experts we spoke with told us this involves consideration of both capital and annual costs, including how expected water savings compare to costs related to energy penalties and other factors. Anticipated future increases in water-related costs could prompt a developer to use a water-saving alternative. For example, a recent report by the Electric Power Research Institute estimates that a power plant's economic trade-offs vary considerably depending on its location and that high water costs could make dry cooling less expensive annually than wet cooling.[23]

The National Energy Technology Laboratory is funding research and development projects aimed at minimizing the drawbacks of advanced cooling technologies and alternative water sources. In 2008, the laboratory awarded close to $9 million to support research and development of projects that, among other things, could improve the performance of dry cooled technologies, recover water used to reduce emissions at coal plants for reuse, and facilitate the use of alternative water sources in cooling towers. Such research endeavors, if successful and deemed economical, could alter the trade-off analysis power plant developers conduct in favor of nontraditional alternatives to cooling.

STATES WE CONTACTED VARY IN THE EXTENT TO WHICH THEY CONSIDER WATER IMPACTS WHEN REVIEWING POWER PLANT DEVELOPMENT PROPOSALS

The seven states that we contacted—Alabama, Arizona, California, Georgia, Illinois, Nevada, and Texas—vary in the extent to which they consider the impacts that power plants will have on water when they review power plant water use proposals. Specifically, these states have differences in water laws that may influence their oversight of power plant water use. Some also have other regulatory policies and requirements specific to power plants and water use. Still other states require additional levels of review that may affect their states' oversight of how power plants use water.

States We Contacted Have Differences in Water Laws that Influence Their Oversight of Water Use by Proposed Power Plants

Differences in water laws in the seven states we contacted—Alabama, Arizona, California, Georgia, Illinois, Nevada, and Texas—influence the steps that power plant developers need to take to obtain approval to use surface or groundwater, and provide for varying levels of regulatory oversight of power plant water use. Table 4 shows the differences in water laws and water permitting for the seven states we contacted.

With regard to surface water—the source of water most often used for power plant cooling nationally—of the seven states we contacted, all but Alabama required power plant developers to obtain water permits through the state agency that regulates the water supply. However, the states requiring permits varied in how the permits were obtained and under what circumstances. For example, in general, under Illinois law, water supply permits are only necessary if the surface water is defined as a public water body, which covers most major navigable lakes, rivers, streams, and waterways as defined by the Illinois Office of Water Resources. However, for any other surface water body, such as smaller rivers and streams, no such permit is required. To obtain a permit to use water in a power plant in Illinois, developers must file an application with the Illinois Office of Water Resources. In determining whether to issue a permit, the Office of Water Resources requires the applicant to address public comments and evaluates USGS streamflow data to determine whether restrictions on water use are needed. In some instances, such as to support fish and other wildlife, the state may designate a minimum level of flow required for a river or stream and restrict the amount of water that can be used by a power plant or other water user when that minimum level is reached. The Director of the Office of Water Resources told us that the office has sometimes encouraged power plant operators to establish backup water sources, such as onsite reservoirs, for use when minimum streamflow levels are reached and water use is restricted. In contrast, under Georgia and Alabama riparian law, landowners have the right to the water on and adjacent to their land, and both states require users who have the capacity to withdraw (Alabama) or actually withdraw (Georgia) an average of more than 100,000 gallons per day to provide information to the state concerning their usage and legal rights to the water. However, this requirement is applied differently in the two states. Alabama requires that water users register their planned water use for informational purposes with the Alabama Office of Water Resources but does not require users to obtain a permit for the water withdrawal or conduct analysis of the impact of the proposed water use.[24] In contrast, Georgia requires water users to apply for and receive a water permit from the Georgia Environmental Protection Division. In determining whether to issue a permit for water use, this Georgia agency analyzes the potential effect of the water use on downstream users and others in the watershed. State water regulators in Georgia told us they have never denied an application for water use in a power plant due to water supply issues since there has historically been adequate available water in the state. For more details on Georgia's process for approving water use in power plants, see appendix IV.

Groundwater laws in the selected states we reviewed also varied and affected the extent to which state regulators provided oversight over power plant water use. In four of the seven states—Alabama, California, Illinois, and Texas—groundwater is largely unregulated at the state level, and landowners may generally freely drill new wells and use groundwater as they wish unless restricted by local entities, such as groundwater conservation districts. However, in three of the seven states we contacted—Arizona, Georgia, and Nevada—state-issued water permits are required for water withdrawals for some or all regions of the state. For example, in Nevada, which has 256 separate groundwater basins, and in which most of the in-state power generation uses groundwater for cooling, state water law follows the doctrine of prior appropriation. A power plant developer or other entity wanting to acquire a new water right for groundwater must apply for a water permit with the Nevada Division of Water Resources. In evaluating the application for a water permit, the Division determines if water is available—referred to as unappropriated; whether the proposed use will conflict with existing water rights or domestic wells; and whether the use of the water is in the public interest. In determining whether groundwater is available, if the Division of Water Resources determines that the amount of water that replenishes the groundwater basin annually is greater than the existing committed ground water rights in a given basin, unappropriated water may be available for appropriation.[25] In two cases where groundwater was being considered for possible power plants, the State Engineer, the official in the Division of Water Resources who approves permits, either denied the application or expressed reservations over the use of groundwater for cooling.[26] For example, in one case, the State Engineer noted that large amounts of water should not be used in a dry state like Nevada when an alternative, like dry cooling, that is less water intensive was available.

In contrast, in Texas, where 8 percent of in state electricity capacity uses groundwater for cooling, state regulators do not issue groundwater use permits or routinely review a power plant or other users' proposed use of the groundwater. Texas groundwater law is based on the "rule of capture," meaning landowners, including developers of power plants that own land, have the right to the water beneath their property. Landowners can pump any amount of water from their land, subject to certain restrictions, regardless of the effect on other wells located on adjacent or other property.[27] Although Texas state water regulators do not issue water permits for the use of groundwater, in more than half the counties in Texas, groundwater is managed locally through groundwater conservation districts which are generally authorized by the Texas Legislature and ratified at the local level to protect groundwater. These districts can impose their own requirements on landowners to protect water resources. This includes requiring a water use permit and, in some districts, placing restrictions on the amount of water used or location of groundwater wells for landowners.[28]

Table 4. State Water Laws and Permit Requirements for Water Supply in Seven Selected States

State	Type of state water laws		State water permit required	
	Surface water	Groundwater	Surface water	Groundwater
Alabama	Riparian	Reasonable use	No[a]	No[a]
Arizona	Prior appropriation	Reasonable use[b]	Yes	Yes[b]
California	Riparian and prior appropriation	Reasonable use and prior appropriation	Yes	No
Georgia	Riparian	Reasonable use	Yes[c]	Yes[c]
Illinois	Other doctrine[d]	Reasonable use	Yes[e]	No
Nevada	Prior appropriation[f]	Prior appropriation[f]	Yes	Yes
Texas	Prior appropriation	Rule of capture	Yes	No[g]

Source: GAO analysis of state laws, documents, and discussions with state officials.

[a] Alabama issues a certificate of use upon registration to users with a capacity to withdraw 100,000 gallons of water per day or more.

[b] Arizona issues state permits for groundwater in areas of severe water overdraft where water shortages could occur, known as Active Management Areas, established under Arizona law. Reasonable use would not apply in these areas.

[c] Georgia issues water permits for users withdrawing more than 100,000 gallons a day.

[d] Illinois surface water law is based on various state statutes.

[e] Illinois issues surface permits only for public water bodies, which excludes some surface water.

[f] In Nevada, water appropriated from either surface or underground sources is limited to that which is reasonably required for beneficial use.

[g] Water use permits can be required locally in Texas through Groundwater Conservation Districts.

States We Contacted Have Other Regulatory Policies That Influence the Extent of Water Use Oversight for Proposed Power Plants

Oversight of water use by proposed power plants in the selected states may be influenced by regulatory policies and requirements that formally emphasize minimizing freshwater use by power plants and other new industrial users. With respect to regulatory policies, of the 7 states, California and Arizona have established formal policies or requirements to encourage power plant developers to consider alternative cooling methods and reduce the amount of freshwater used in a proposed power plant. Specifically:

- California, a state that has faced constrained water supplies for many years, established a formal policy in 1975 that requires applicants seeking to use water in power plants to consider alternative water sources before proposing the use of freshwater.[29] More recently, the California Energy Commission, the state agency that is to review and approve power plant developer applications, reiterated in its 2003 Integrated Energy Policy Report, the 1975 policy that the commission would only approve power plants using freshwater for cooling in limited circumstances.[30] Furthermore, state regulators at the Commission told us that in discussing potential new power plant developer applications, commission staff encourage power plant developers to consider using advanced cooling technologies, such as dry cooling or alternative water sources, such as effluent from sewage treatment plants. Between January 2004 and April 2009, California regulators approved 10 thermoelectric power plants—3 that will use dry cooling; 6 that will use an alternative water source, such as reclaimed water; and 2 that will use freshwater purchased from a water supplier, such as a municipal water district, for power plant cooling.[31] Of 20 additional thermoelectric power plant applications pending California Energy Commission approval, developers have proposed 11 plants that plan to use dry cooling, 8 plants that plan to use an alternative water source, and 1 that plans to use freshwater for cooling.[32] For more details on California's process for approving water use in power plants, see appendix III.
- In Arizona, where there is limited available surface water and where groundwater is commonly used for power plant cooling, the state has requirements to minimize how much water may be used by power plants. Specifically, in Active Management Areas—areas the state has determined require regulatory oversight over the use of groundwater—the state requires that developers of new power plants 25 megawatts or larger using groundwater in a wet recirculating system with a cooling tower, design the plants to reuse the cooling water to a greater extent than what is common in the industry. Plants must cycle water through the cooling loop at least 15 times before discharging it, whereas, according to an Arizona public utility official, outside of Active Management Areas plants would generally cycle water 3 to 7 times.[33] These additional cycles result in water savings, since less water must be withdrawn from ground or surface water sources to replace discharges, but can require plant operators to undertake more costly and extensive treatment of the cooling water and to more carefully manage the plant cooling equipment to avoid mineral buildup.[34] Arizona officials also told us they encourage the use of alternative water sources for

cooling and have informally encouraged developers to consider dry cooling. According to Arizona state officials, no plants with dry cooling have been approved to date in the state and, due mostly to climatic conditions, dry cooling is probably too inefficient and costly to currently be a viable option. For details on Arizona's process for approving water use in power plants, see appendix II.

In contrast to California and Arizona, water supply and public utility commission officials in the other 5 selected states told us their states had not developed official state policies regarding water use by power plants. For example, Alabama, a state where water has traditionally been plentiful, has not developed a specific policy related to power plant water use or required the use of advanced cooling technologies or alternative water sources. Additionally, the state does not require that power plant developers and other proposed water users seek a water use permit; rather power plant operators are only required to register their maximum and average expected water use with the state and report annual usage. State officials told us that they require this information so that they can know how much water is being used but that their review of power plant water use is limited. Officials from the state's Public Service Commission, responsible for certifying the development of power plants, said their office does not have authority to regulate a utility's water use and, therefore, generally does not analyze how a proposed power plant will affect the water supply. Rather, their office focuses on the reasonableness of power plant costs.[35]

Similarly, Illinois, where most power plants use surface water for cooling and water is relatively plentiful, has not developed a policy on water use by thermoelectric power plants or required the use of advanced cooling technologies or alternative water sources, according to an official at the Office of Water Resources. However, the Illinois Office of Water Resources does require power plant operators, like other proposed water users, to apply for water permits for use of surface water from the major public water bodies.

States We Contacted May Require Additional Levels of Review That Affect Oversight

Three of the states we selected—Arizona, Nevada, and California—conduct regulatory proceedings that consider water availability, in addition to determining whether to issue a water permit, while the other states do not. In Arizona, water use for power plants is subject to three reviews: (1) the process for a prospective water user to obtain a water permit, if required; (2) review by a committee of the Arizona Corporation Commission, known as the Arizona Power Plant and Transmission Line Siting Committee; and (3) review by the Commission as part of an overall evaluation of the plant's feasibility and its potential environmental and economic impacts. Both the Committee and Commission evaluate water supply concerns, along with other environmental issues, and determine whether to recommend (Committee) or issue (Commission) a Certificate of Environmental Compatibility, which is necessary for the plant to be approved.[36] Water supply concerns have been a factor in denying such a certificate for a proposed power plant. For example, in 2001, the Commission denied an application to build a new plant over concerns that groundwater withdrawals for cooling water would not be naturally replenished and, thereby, would reduce surface water availability which could adversely affect the habitat for an endangered species. For more details on Arizona's processes for approving water use in power plants see appendix II.

Similarly, in Nevada and California, several state agencies may play a role in the approval of water use and the type of cooling technology used by power plants. In Nevada, although water permits for groundwater and surface water are issued by the State Engineer, the Public Utilities Commission oversees final power plant approval under the Utility Environmental Protection Act. Even if the power plant developer has obtained a water permit, water use could play a role in the review process if the plant's use of the cooling water or technologies has environmental effects that need to be mitigated. Additionally, as in a number of states where electricity rates are regulated, the Public Utilities Commission could consider the effect of dry cooling on electricity rates. In California, the California Energy Commission reviews all aspects of power plant certifications, including issuing any water permits and approvals for cooling technologies.[37] According to a California Energy Commission official, during this process the Commission works with other state and local agencies to ensure their requirements are met.

The other four states we contacted do not conduct reviews of how power plants will affect water availability beyond issuing a water use permit or certificate of registration. Public utility regulators in Illinois, Texas, Alabama, and Georgia told us they had no direct role in regulating water use or cooling technologies in power plants. Officials from the Public Utility Commission of Texas noted that since they do not regulate electricity rates in most of the state, the Commission plays no role in the approval of power plants in most areas. In other areas, they told us water use and cooling technologies were not reviewed by the Commission. Similarly, in Illinois—a state that does not regulate electricity rates—an official from the Illinois Commerce Commission stated that the agency had no role in reviewing water use or cooling technologies for power plants. While Georgia and Alabama are states that regulate electricity rates, officials from their Public Service Commissions—the state agencies regulating electricity rates—noted that they focus on economic considerations of power generation and not the impact that a power plant might have on the state's water supply.

SOME FEDERAL WATER DATA ARE USEFUL FOR EVALUATING POWER PLANT APPLICATIONS, BUT LIMITATIONS IN OTHER FEDERAL DATA MAKE THE IDENTIFICATION OF CERTAIN WATER USE TRENDS MORE DIFFICULT

State water regulators rely on data on water availability collected by USGS's streamflow gauges and groundwater studies and monitoring stations when they are evaluating developers' proposals for new power plants. In contrast, state water regulators do not routinely rely on federal data on water use when evaluating power plant applications, although these data are used by water and industry experts, federal agencies, and others to analyze trends in the industry. However, these users of federal data on water use identified a number of limitations with the data that they believe limits its usefulness.

State Water Regulators and Others Rely on Federal Data on Water Availability to Evaluate Power Plant Proposals

State water regulators, federal agency officials, and water experts we spoke with agreed that federal data on water availability are important for multiple purposes, including for deciding whether to approve power plant developer proposals for water permits and water rights. Most state water regulators we contacted explained that they rely upon federal data on water availability, particularly streamflow and groundwater data collected by USGS, for permitting decisions and said these data helped promote more informed water planning. For example, water regulatory officials from the Texas Commission on Environmental Quality—the agency that evaluates surface water rights applications from prospective water users in Texas—told us that streamflow data collected by USGS are a primary data source for their water model that predicts how water use by power plants and others applying for water rights will impact state water supplies and existing rights holders.

USGS's network of streamflow gauges and groundwater monitoring stations provide the only national data of their kind on water availability over long periods. As a result, state officials told us that these data are instrumental in predicting how much water is likely to be available in a river under a variety of weather conditions, such as droughts. For example, state regulators in Georgia and Illinois told us that they rely on USGS streamflow data to determine whether or not to establish special conditions on water withdrawal permits, such as minimum river flow requirements that affect the amount of cooling water a power plant can withdraw during periods when water levels in the river are low. State water regulators in Nevada also told us they rely on a number of data sources, including USGS groundwater studies, to determine the amount of time necessary for water to naturally refill a groundwater basin. This information helps them ensure that water withdrawals for power plants and others are sustainable and do not risk depleting a groundwater basin.

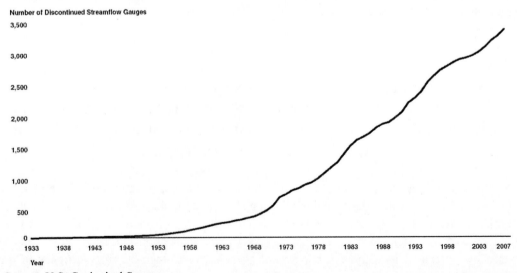

Source: U.S. Geological Survey.

Figure 8. Cumulative Number of Discontinued U.S. Geological Survey Streamflow Gauges with 30 or More Years of Record, 1933-2007.

Table 5. Water Data Considered in Support of State Water Regulators' Permitting Decisions

	USG S data on water availability		State, industry, academic, or other data
	Groundwater	**Streamflow**	
Alabama	a	a	a
Arizona	b	b	b
California	c	Yes	Yes
Georgia	Yes	Yes	Yes
Illinois	c	Yes	Yes
Nevada	Yes	Yes	Yes
Texas	c	Yes	Yes

Source: GAO analysis of information provided by state regulators.

[a] Alabama officials told us they are not authorized to issue water withdrawal permits and, thus, do not rely on USGS water availability data for this purpose. However they rely on these data for a variety of other purposes.

[b] Arizona officials told us that, in practice, they do not often rely on USGS streamflow data for permitting because surface water is fully allocated throughout the state. Similarly, groundwater availability data is not routinely relied upon for permits for groundwater rights in Active Management Areas, since most power plant developers purchase existing rights, rather than apply for a new right. Outside of Active Management Areas, water users only seek drilling permits, which requires limited review. However, surface and groundwater availability data may be relied on to support the Line Siting Committee and the Arizona Corporation Commission's decision to issue a Certificate of Environmental Compatibility.

[c] These states do not issue permits for groundwater at the state level. However, in California, any groundwater use for a power plant would be permitted, if necessary, through the California Energy Commission, which regulates the licensing of power plants.

State regulators told us that while federal water availability data is a key input into their decisionmaking process for power plant permits, they also rely on a number of other sources of data, as shown in table 5. These include data that they themselves collect and data collected by universities; private industry, such as power plant developers; and various other water experts.

Some state regulators and water experts we spoke with expressed concern about streamflow gauges being discontinued, which they said may make evaluating trends in water availability and water planning more difficult in the future. Without accurate data on water availability, decisions about water planning and allocation of water resources—including power plant permitting decisions—may be less informed, according to regulators and experts. For example, an official from Arizona told us that a reduction in streamflow gauges would adversely impact the quality of the states' water programs and that state budget constraints have made it increasingly difficult to allocate the necessary state funds to ensure cooperatively-funded streamflow gauges remain operational. Similarly, an official from the Texas Commission on Environmental Quality told us that if particular streamflow gauges were discontinued, water availability records would be unavailable to update existing data for their water availability models—which are relied upon for water planning and permitting decisions—and alternative data would be needed to replace these missing data. USGS officials told us that the cumulative number of streamflow gauges with 30 or more years of record that have been discontinued has increased, as seen in figure 8, due to budget constraints.

Water Experts, Federal Agencies, and Others Value Federal Data on Water Use for Analyzing Industry Trends but Identified Limitations In These Data

Unlike federal data on water availability, federal data on water use is not routinely relied upon by state officials we spoke with to make regulatory decisions; but, instead is used by a variety of data users to identify trends in the industry. Specifically, data users we spoke with, including water experts, representatives of an environmental group, and federal agency officials, identified the following benefits of the water use data collected by USGS and EIA:

- *USGS Data on Water Use.* A number of users of federal water data we spoke with told us that USGS's 5-year data on thermoelectric power plant water use are the only centralized source of long-term, national data for comparing water use trends across sectors, including for thermoelectric power plants. As a result, they are valuable data for informing policymakers and the public about the state of water resources, including changes to water use among power plants and other sectors. For example, one utility representative we spoke with said that USGS data are important for educating the public about how power plants use water and the fact that while thermoelectric power plants withdraw large amounts of water overall—39 percent of U.S. freshwater withdrawals in 2000—their water consumption as an industry has been low—3 percent of U.S. freshwater consumption in 1995. Furthermore, some state water regulators told us that USGS's water use data allow them to compare their state's water use to that of other states and better evaluate and plan around their state's water conditions.[38] An Arizona Department of Water Resources official, for example, told us that USGS's water use data are essential for understanding how water is used in certain parts of the state where the Department has no ability to collect such data.[39]
- *EIA Data on Water Use.* EIA's annual data are the only federally-collected, national data available on water use and cooling technologies at individual power plants; and data users noted that EIA's national data were useful for analyzing the water use characteristics of individual plants, as well as for comparing water use across different cooling technologies. For example, officials at USGS and the National Energy Technology Laboratory told us that they use EIA data to research trends in current and future thermoelectric power plant and other categories of water use. Specifically, USGS utilizes EIA's data on individual plant water use, in addition to data from state water regulators and individual power plants, to develop county and national estimates of thermoelectric power plant water use. USGS officials explained that in some of their state offices, such as California and Texas, agency staff primarily use EIA and other federal data to develop USGS's 5-year thermoelectric power plant water use estimates. Officials from USGS also explained that other USGS state offices use EIA data on water use to corroborate their estimates of thermoelectric power plant water withdrawals and to identify the cooling technology

utilized by power plants. Similarly, officials at the National Energy Technology Laboratory have extensively used EIA's data on individual power plant water withdrawals and consumption to develop estimates of how freshwater use by thermoelectric power plants will change from 2005 to 2030.

However, data users we spoke with also identified a number of shortcomings in the federal data on water use, collected by USGS and EIA, that limits their ability to conduct certain types of industry analyses and understanding of industry trends. Specifically, they identified the following issues, along with others that are detailed in appendix V.

- *Lack of comprehensive data on the use of advanced cooling technologies.* Currently, EIA does not systematically collect information on power plants' use of advanced cooling technologies. In the EIA database, for example, data on power plants' use of advanced cooling technologies is incomplete and inconsistent—not all power plants report information on their use of advanced cooling technologies or do so in a consistent way. Lacking these national data, it is not possible without significant additional work to comprehensively identify how many power plants are using advanced cooling technologies, where they are located, and to what extent the use of these technologies has reduced the use of freshwater. According to a study by the Electric Power Research Institute, although the total number of dry cooled plants is still small relative to plants using traditional cooling systems, the use of advanced cooling technologies is becoming increasingly common.[40] As these technologies become more prevalent, we believe that information about their adoption would help policymakers better understand the extent to which advanced cooling technologies have been successful in reducing freshwater use by power plants and identify those areas of the country where further adoption of these technologies could be encouraged. EIA officials told us they formally coordinate with a group of selected stakeholders every 3 years to determine what changes are needed to EIA data collection forms. They told us they have not previously collected data on advanced cooling technologies because EIA's stakeholder consultation process had not identified these as needed data. However, these officials acknowledged that EIA has not included USGS as a stakeholder during this consultation process and were unaware of USGS' extensive use of their data. In discussing these concerns, EIA officials also said that they did not expect that collecting this information would be too difficult and agreed that such data could benefit various environmental and efficiency analyses conducted by other federal agencies and water and industry experts. Furthermore, in discussing our preliminary findings, EIA officials also said they believed that EIA could collect these data during its triennial review process by, for example, adding a reporting code for these types of cooling systems. However, they noted that they would have to begin the process soon to incorporate it into their ongoing review.
- *Lack of comprehensive data on the use of alternative water sources.* Our review of federal data sources indicates that they cannot be used to comprehensively identify plants using alternative water sources. EIA routinely reports data on individual plant water sources, but we found that these data do not always identify whether the source of water is an alternative source or not. Similarly, while the USGS data identify thermoelectric power plants using ground, surface, fresh, and saline water, they do

not identify those using alternative water sources, such as reclaimed water. While a goal of USGS's water use program is to document trends in U.S. water use and provide information needed to understand the nation's water resources, USGS officials said budget constraints have limited the water use data the agency can provide, and has led to USGS discontinuing distribution of data on one alternative water source—reclaimed water. According to two studies we reviewed, use of some alternative water sources is becoming more common and, based on our discussions with regulators and power plant developers, there is much interest in this nonfreshwater option, particularly in areas where freshwater is constrained. As use of these alternative water sources becomes more prevalent, we believe that information about how many plants are using these resources and in what locations, could help policymakers better understand how the use of alternative water sources by power plants can replace freshwater use and help identify those areas of the country where such substitution could be further encouraged.

- *Incomplete water and cooling system data.* Though part of EIA's mission is to provide data that promote public understanding of energy's interaction with the environment, EIA does not collect data on the water use and cooling systems of two significant components of the thermoelectric power plant sector. First, in 2002, EIA discontinued its reporting of water use and cooling technology information for nuclear plants. According to data users we spoke with, this is a significant limitation in the federal data on water use and makes it more difficult for them to monitor trends in the industry. For example, USGS officials said that the lack of these data make developing their estimates for thermoelectric power plant water use more difficult because they either have to use older data or call plants directly for this information, which is resource intensive. EIA officials told us they discontinued collection of data from nuclear plants due to priorities stemming from budget limitations.[41] Second, EIA does not collect water use and cooling system data from operators of some combined cycle thermoelectric power plants. Combined cycle plants represented about 25 percent of thermoelectric capacity in 2007, and constituted the majority of thermoelectric generating units built from 2000 to 2007. According to EIA officials, water use and cooling technology data are not collected from operators of combined cycle plants that are not equipped with duct burning technology—a technology that injects fuel into the exhaust stream from the combustion turbine to provide supplemental heat to the steam component of the plant. However, these plants use a cooling system and water, as do other combined cycle and thermoelectric power plants whose operators are required to report to the agency. As a result, data EIA currently collects on water use and cooling systems for thermoelectric power plants is incomplete. EIA officials acknowledged that not collecting these data results in an incomplete understanding of water use by these thermoelectric power plants; however, budget limitations have thus far precluded collection of such data. According to a senior EIA staff in the Electric Power Division, since speaking with GAO, the agency has begun exploring options for collecting these data as part of its current data review process.

- *Discontinued distribution of thermoelectric power plant water consumption data.* One of the stated goals of USGS's water use program is to document trends in U.S. water use, but officials told us that a lack of funding has prompted the agency to discontinue distribution of data on water consumption for thermoelectric power plants and other water users.[42] These USGS officials told us they would like to restart distribution of the data on water consumption by thermoelectric power plants and other water users if additional funding were made available, because such data can be used to determine the amount of water available for reuse by others. Similarly, some users of federal water data told us that not having USGS data on consumption limits their and the public's understanding of how power plant water consumption is changing over time, in comparison to other sectors. They said that the increased use of wet recirculating technologies, which directly consume more water but withdraw significantly less than once-through cooling systems, has changed thermoelectric power plant water use patterns.[43]

In a 2002 report, the National Research Council recommended that USGS's water use program be elevated from one of water use accounting to water science—research and analysis to improve understanding of how human behavior affects patterns of water use.[44] Furthermore, the council's report concluded that statistical analysis of explanatory variables, like cooling system type or water law, is a promising technique for helping determine patterns in thermoelectric power plant water use. The report suggested these and other approaches could help USGS improve the quality of its water use estimates and the value of the water data it reports. USGS has proposed a national water assessment with the goal of, among other things, addressing some of the recommendations made by the National Research Council report. USGS officials also told us such an initiative would make addressing some of the limitations in USGS water use data identified by water experts and others possible, such as reporting data on water consumption and by hydrologic code.

CONCLUSIONS

While much of the authority for regulating water use resides at the state level, the federal government plays an important role in collecting and distributing information about water availability and water use across the country that can help promote more effective management of water resources. However, the lack of collection and reporting of some key data related to power plant water use limits the ability of federal agencies and industry analysts to assess important trends in water use by power plants, compare them to other sectors, and identify the adoption of new technologies that can reduce freshwater use. Without this comprehensive information, policymakers have an incomplete picture of the impact that thermoelectric power plants will have on water resources in different regions of the country and will be less able to determine what additional activities they should encourage for water conservation in these areas. Moreover, although both EIA and USGS seek to provide timely and accurate information about the electricity sector's water use, they have not routinely coordinated their efforts in a consistent and formal way. As a result, key water data collected by EIA and used by USGS have been discontinued or omitted and important trends in the electricity sector have been overlooked. EIA's ongoing triennial review of the data it collects about power plants and the recent passage of the Secure Water Act, that authorizes funding for USGS to report data on water use to Congress, provide a timely opportunity to address gaps in federal data collection and reporting and improve coordination between USGS and EIA in a cost-effective way.

RECOMMENDATIONS FOR EXECUTIVE ACTION

We are making seven recommendations. Specifically, to improve the usefulness of the data collected by EIA and better inform the nation's understanding of power plant water use and how it affects water availability, we recommend that the Administrator of EIA consider taking the following four actions as part of its ongoing review of the data it collects about power plants:

- add cooling technology reporting codes for alternative cooling technologies, such as dry and hybrid cooling, or take equivalent steps to ensure these cooling technologies can be identified in EIA's database;
- expand reporting of water use and cooling technology data to include all significant types of thermoelectric power plants, particularly by reinstating data collection for nuclear plants and initiating collection of data for all combined cycle natural gas plants;
- collect and report data on the use of alternative water sources, such as treated effluent and groundwater that is not suitable for drinking or irrigation, by individual power plants; and
- include USGS and other key users of power plant water use and cooling system data as part of EIA's triennial review process.

To improve the usefulness of the data collected by USGS and better inform the nation's understanding of power plant water use and how it affects water availability, we recommend that the Secretary of the Interior consider:

- expanding efforts to disseminate available data on the use of alternative water sources, such as treated effluent and groundwater that is not suitable for drinking or irrigation, by thermoelectric power plants, to the extent that this information becomes available from EIA; and
- reinstating collection and distribution of water consumption data at thermoelectric power plants.

To improve the overall quality of data collected on water use from power plants, we recommend that EIA and USGS establish a process for regularly coordinating with each other, water and electricity industry experts, environmental groups, academics, and other federal agencies, to identify and implement steps to improve data collection and dissemination.

APPENDIX I. OBJECTIVES, SCOPE AND METHODOLOGY

At the request of the Chairman of the House Committee on Science and Technology, we reviewed (1) technologies and other approaches that can help reduce freshwater use by power plants and what, if any, drawbacks there are to implementation; (2) the extent to which selected states consider water impacts of power plants when reviewing power plant development proposals; and (3) the usefulness of federal water data to experts and state regulators who evaluate power plant development proposals. We focused our evaluation on thermoelectric power plants, such as nuclear, coal, and natural gas plants using a steam cycle. We did not consider the water supply issues associated with hydroelectric power, since the process through which these plants use water is substantially different from that of thermoelectric plants (e.g., water is used as it passes through a dam but is not directly consumed in the process). We also focused the review on water used during the production of electricity at power plants, and did not include water issues associated with extracting fuels used to produce electricity.

To understand technologies and other approaches that can help reduce freshwater use by power plants and their drawbacks, we reviewed industry, federal, and academic studies on advanced cooling technologies and alternative water sources that discussed their benefits, such as reduced freshwater use, and what, if any, drawbacks their implementation entails. These included studies with information on power plants' use of water and the drawbacks of nonfreshwater alternatives conducted by the Electric Power Research Institute, the Department of Energy's National Energy Technology Laboratory, and others. We discussed these trade-offs with various experts, including power plant and cooling system manufacturers, such as GEA Power Cooling Inc., General Electric, Siemens, and SPX Cooling Technologies; other industry groups and consultants, such as the Electric Power Research Institute, Maulbetsch Consulting, Nalco, and Tetra Tech; an engineering firm, Black & Veatch; and federal, national laboratory, and academic sources. To get a user perspective on these different technologies and alternative water sources, we met with power plant operators, including Arizona Public Service Company, Calpine, Georgia Power Company, and Sempra Generation. We also spoke with representatives from and reviewed reports prepared by other National Laboratories, such as the Department of Energy's Argonne National Laboratory, to understand related research activities concerning water and electricity. To better understand how the differences in cooling technologies and heat sources used by power plants affect power plant configuration and design, we toured three power plant facilities in Texas—Comanche Peak (nuclear, once-through cooling), Limestone (coal, wet recirculating with cooling towers), and Midlothian (natural gas combined cycle, dry cooling).

To determine the extent to which selected states consider water impacts of power plants when reviewing power plant development proposals, we conducted case study reviews of three states Arizona, California, and Georgia. These states were selected because of their historic differences in water availability, differences in water law, high energy production, and large population centers. We did not attempt to determine whether states' efforts were reasonable or effective, rather we only described what states do to consider water impacts when making power plant siting decisions. For each of these case study states, we met with state water regulators and power plant developers to understand how water planning and permitting decisions are approached from both a regulatory and private industry perspective. We also met with water research institutions and other subject matter experts to understand current and future research related to water impacts of power plants and the extent to which these research endeavors help inform power plant development proposals and regulatory water permitting decisions. Specifically, in California we met with the California Department of Water Resources; the California Energy Commission; the California State Water Resources Control Board; the San Francisco Bay Regional Water Quality Control Board; and the U.S. Geological Survey's (USGS) California Water Science Center. In Georgia we met with the Georgia Environmental Protection Division; the Georgia Public Service Commission; the Georgia Water Resources Institute; the Metropolitan North Georgia Water Planning District; the U.S. Army Corps of Engineers, South Atlantic Division; and the USGS Georgia Water Science Center. In Arizona we met with the Arizona Corporation Commission; the Arizona Department of Environmental Quality; the Arizona Department of Water Resources; the Arizona Power Plant and Transmission Line Siting Committee; the Arizona Office of Energy, Department of Commerce; the Arizona Water Institute, and the USGS Arizona Water Science Center. In addition, we reviewed state water laws and policies for thermoelectric power plant water use, selected power plant operator proposals to use water, and state water regulators' water permitting decisions. We also reviewed selected public utility commission dockets and testimonies describing various power plant siting decisions to understand what, if any, water issues were addressed. To broaden our understanding of how states consider the water impacts of power plants when reviewing power plant development proposals, we supplemented our case studies by conducting interviews and reviewing documents from four additional states Nevada and Alabama—which shared watersheds with the case study states—and Illinois and Texas, which are large electricity producing states with sizable population centers. For each of these four states, we spoke with the primary state water regulatory agencies—the Alabama Office of Water Resources, the Illinois Office of Water Resources, the Nevada Division of Water Resources, and the Texas Commission on Environmental Quality—to understand how state water regulators consider the impacts of power plant operators' proposals to use water. In Texas, additional discussions were held with the Public Utility Commission of Texas; the Texas Water Development Board; the University of Texas; and the USGS Texas Water Science Center to further understand how water supply issues and energy demand are managed in Texas. In Alabama, we held additional discussions with officials from the Alabama Public Service Commission and the Alabama Department of Environmental Management to learn more about how Alabama's state water regulators and power plant operators manage water supply and energy demand. In Nevada, we held a discussion with an official from the Public Utilities Commission of Nevada to determine how they evaluate cooling technologies and water issues in plant siting certification proceedings. We also contacted the Illinois Commerce Commission.

Finally, to determine how useful federal water data are to experts and state regulators who evaluate power plant development proposals, we reviewed data and analysis from the Energy Information Administration (EIA), USGS, and the Department of Energy's National Energy Technology Laboratory and analyzed how the data were being used. We also conducted interviews with federal agencies, including the Bureau of Reclamation; EIA; Environmental Protection Agency; Tennessee Valley Authority; U.S. Army Corps of Engineers; and USGS to understand whether each organization also collected water data and their opinions about the strengths and limitations of EIA and USGS data. We spoke with several regional offices for the Bureau of Reclamation, including the Lower Colorado and Mid-Pacific offices to understand federal water issues in California, Arizona, and Nevada. In addition, to understand how valuable federal water data are to experts and state regulators who evaluate power plant development proposals to use water, we conducted interviews and reviewed documents from state water regulators and public utility commissions, as well as water and electricity experts at environmental and water organizations, such as the Pacific Institute and Environmental Defense Fund; at universities such as the Georgia Institute of Technology; Southern Illinois University, Carbondale; and the University of Maryland, Baltimore County; and experts from industry, national laboratories, and other organizations and universities previously mentioned. We also contacted other electricity groups, including the North American Electric Reliability Corporation and the National Association of Regulatory Utility Commissioners, to get a broader understanding of how the electricity industry addresses water supply issues.

We conducted this performance audit from October 2008 through October 2009, in accordance with generally accepted government auditing standards. Those standards require that we plan and perform the audit to obtain sufficient, appropriate evidence to provide a reasonable basis for our findings and conclusions based on our audit objectives. We believe that the evidence obtained provides a reasonable basis for our findings and conclusions based on our audit objectives.

APPENDIX II. REVIEW OF PROPOSALS TO USE WATER IN NEW POWER PLANTS IN ARIZONA

Background

Arizona, with a population of 6.5 million, was the 16th most populous state in the country in 2008 and was one of the fastest growing states, growing at a rate of 2.3 percent from 2007 to 2008. Most of the land in Arizona is relatively dry, therefore, water for electricity production is limited. For 2007, Arizona accounted for 2.7 percent of U.S. net electricity generation, ranking it 13th, with most generation coming from coal (36 percent); natural gas (34 percent); nuclear (24 percent); and renewable sources, such as hydroelectric (6 percent), although the state has a strong interest in developing solar and other renewable sources.

Arizona Water Law and Policy

Arizona relies on three water sources for electricity production: (1) surface water, including the Colorado River; (2) groundwater; and (3) effluent. Arizona water law varies depending on the source and the user's location, specifically:

- *Surface water.* The use of surface water in Arizona is determined by the doctrine of prior appropriation. The Arizona Department of Water Resources issues permits to use surface water statewide, with the exception of water from the Colorado River.[45]

The federal government developed water storage and distribution via a series of canals to divert water from the Colorado River to southern Arizona, and the Bureau of Reclamation issues contracts for any new water entitlements related to Colorado River water, in consultation with the Arizona Department of Water Resources.

- *Groundwater.* The use of groundwater depends on its location. Because some areas receive seasonal rain and snow, average annual precipitation can vary by location, from 3 to over 36 inches of moisture. The state established five regions where groundwater is most limited known as Active Management Areas. Permits to use groundwater in these five areas are coordinated through the Arizona Department of Water Resources, which provides several permitting options for power plants.[46] Outside Active Management Areas, the state subjects groundwater to little regulation or monitoring and generally only requires users to submit a well application to the Department of Water Resources.
- *Effluent.* Effluent is owned by the entity that generates it until it is discharged into a surface water channel. The owner has the right to put effluent to beneficial use or convey it to another entity, such as a power plant, that will put it to beneficial use. However, once it is discharged from the pipe, generally into a surface water body, such as a river, it is considered abandoned and subject to laws governing surface water.

Arizona has no overall statewide policy on the use of water in thermoelectric power plants. However, in Active Management Areas, the state requires developers of newer power plants with a generating capacity of 25 megawatts or larger to use groundwater in a wet recirculating system with a cooling tower and to cycle water through the cooling loop at least 15 times before discharging it.[47] An official of an Arizona public utility noted that it was more common to cycle water 3 to 7 times outside of Active Management Areas.

Certification and Water Permitting for New Power Plants

Before a power plant developer can begin constructing a power plant with a generating capacity of 100 megawatts or larger, it must go through a two-step certification process and a permitting process, as follows:[48]

- The first step of the certification process involves public hearings before the Arizona Power Plant and Transmission Line Siting Committee, made up of representatives from five state agencies and six additional members appointed by the Arizona Corporation Commission.[49] Although the Line Siting Committee is not required to evaluate water use unless the plant will be located within an Active Management Area, it typically considers water rights, water availability for the life of the power plant, and the environmental effects of groundwater pumping around the plant. Committee members told us they often ask about the planned water sources and whether alternative water sources and cooling technologies are available. If the plant will be located within an Active Management Area, a representative of the Department of Water Resources serving on the Committee takes the lead in

evaluating the plant's potential adverse impacts on the water source, including reviewing state data or U.S. Geological Survey (USGS) studies that document the status and health of the proposed water source. A representative from the Arizona Department of Environmental Quality serving on the Committee considers the plant's potential adverse effects on water quality. Based on this information, as well as the proposed plant's feasibility and its potential environmental and economic impacts, the Committee issues a recommended Certificate of Environmental Compatibility, if appropriate.
- In the second step of the certification process, the Arizona Corporation Commission reviews the power plant developer's application to ensure there is a balance between the state's need for energy and the plant's cost and potential environmental impacts, including water quality, water supply, ecological, and wetlands impacts. The Commission can accept, deny, or modify the Certificate of Environmental Compatibility that was recommended by the Line Siting Committee and has denied some certificates. The Commission places the burden on the applicant to demonstrate that the proposed water supply is sustainable and how any water quality impacts will be mitigated. The Commission does not collect or review additional water data or conduct quality checks on the data provided by the power plant developers.
- The permitting process applies to both water supply and water quality. With respect to water supply, when required, power plant developers who plan to use surface water in most areas of the state or groundwater in an Active Management Area must obtain a water use permit from the Department of Water Resources. When applying for a permit, power plant developers are required to provide information on the amount of water they will use, the source, points of diversion and release, and how the power they generate will be used. For groundwater in an Active Management Area, users are strictly limited to a total volume of water permitted for withdrawal and are subject to annual reporting and an analysis of the impact on other wells. According to an official at the Department of Water Resources, the Department has extensive data on available groundwater for each Active Management Area to assist in determining the effects of groundwater use. With respect to water quality, power plant developers must obtain permits which regulate water quality through the Department of Environmental Quality. Further, power plants discharging into federally-regulated waters also need a National Pollutant Discharge Elimination System permit that covers effluent limitations and sets discharge requirements. This program is intended to ensure that discharges to surface waters do not adversely affect the quality and beneficial uses of such water.

Recent State Decisions about Power Plant Water Use

Between January 2004 and July 2009, Arizona has approved three new power plants, two of which are simple cycle natural gas plants that do not need water for cooling. The third plant is a concentrating solar thermal plant using a wet recirculating system with cooling towers. According to an official from the Arizona Department or Water Resources, once the plant begins operating, it will use 3,000 acre feet of water annually from groundwater and surface water, under contract from an Irrigation District.

Between 1999 and 2002, a large number of applications for power plants in Arizona were filed, most of which were approved.[50] However, at least one plant was denied a Certificate of Environmental Compatibility due to a water supply concern—the potential loss of habitat for an endangered species from possible groundwater depletion. Approved plants used a variety of water sources for cooling, including recycled wastewater, surface water through arrangements with the Central Arizona Project, and groundwater—both directly used or from conversion of agricultural land. No dry cooled power plants have been approved in Arizona, according to state officials. State officials told us dry cooling is too inefficient and costly, but that it may be considered in the future if water shortages become more acute.

APPENDIX III. REVIEW OF PROPOSALS TO USE WATER IN NEW POWER PLANTS IN CALIFORNIA

Background

As of January 2009, California had the nation's largest population—an estimated 38.3 million people—and grew at a rate of 1.1 percent annually from 2008 to 2009. California has significant variations in water availability, with a long coastline; several large rivers, particularly in the north; mountainous areas that receive substantial snowfall; and arid regions, particularly the Mojave Desert in southeastern California. Statewide, California averages 21.4 inches of rain annually, but has suffered significant droughts for the past three years. For 2007, California accounted for 5.1 percent of U.S. net electricity generation, ranking it 4th nationally. California generates electricity primarily from natural gas (55 percent); nuclear (17 percent); and renewable energy sources—primarily hydroelectric, wind, solar, and geothermal (25 percent). California imports 27 percent of its electricity from other states.

California Water Law and Policy

California water law depends on whether the water is surface water or groundwater, specifically:

- *Surface water.* The use of surface water is subject to both the riparian and appropriative rights doctrines. No permit is needed to act upon riparian surface water rights, which result from ownership of land bordering a water source, and are senior to most appropriative rights. Appropriative rights, on the other hand, must be acquired through the State Water Resources Control Board. Applicants for appropriative rights must show, among other things, that the water will be put to beneficial use.
- *Groundwater.* The majority of California's groundwater is unregulated.[51] Additionally, California does not have a comprehensive groundwater permit process

in place, except for groundwater that flows through subterranean streams, which is permitted by the State Water Resources Control Board.

California has several policies that directly and indirectly address how thermoelectric power plants can use water. Specifically:

- California's State Water Resources Control Board, as the designated state water pollution control agency and issuer of surface water rights, established a policy in 1975 that states that the use of fresh inland waters for power plant cooling will only be approved when it is demonstrated that the use of other water supply sources or other methods of cooling would be environmentally undesirable or economically unsound. Freshwater should be considered the last resort for power plant cooling in California. Since that time, according to officials we spoke with, the Board has encouraged the use of alternative sources of cooling water and alternative cooling technologies.
- The California Energy Commission (CEC), the state's principal energy policy and planning organization, in 2003, reiterated the 1975 policy and further required developers to consider whether zero-liquid discharge technologies should be used to reduce water use unless it can be shown that the use of these technologies would be environmentally undesirable or economically unsound. Under these policies, dry cooling and use of alternative water for cooling would be the preferred alternatives.
- The State Water Resources Control Board discourages the use of once-through cooling in power plants due to potential harm to aquatic organisms. The agency is considering a state policy to require power plants using this technology to begin using other cooling technologies or retire from service.

Certification and Water Permitting for New Power Plants

California has a centralized permitting process for new large power plants, including thermoelectric power plants. Developers constructing new power plants with a generating capacity of 50 megawatts or larger must apply for certification with the CEC, the lead state agency for ensuring proposed plants meet requirements of the California Environmental Quality Act and generally overseeing the siting of new power plants.[52] The CEC coordinates review of other state environmental agencies, such as the State Water Resources Control Board and issues all required state permits (air permits, water permits, etc.). Prior to issuing the permits needed to construct a new power plant, the CEC conducts an independent assessment, with public participation, of each proposed plant's environmental impacts; public health and safety impacts; and compliance with federal, state, and local laws, ordinances, and regulations. As part of its review, CEC staff analyze the effect on other water users of power plant developers' proposed use of water for cooling and other purposes, access to needed water supplies throughout the life of the plant, and the plant's impact on the proposed water source and the state's water supply overall.[53] The CEC also ensures power plant developers have obtained the required water supply agreements; analyzed the feasibility of alternative water sources and cooling technologies; and addressed water supply, water quality, and wastewater disposal impacts. The CEC may require implementation of various measures to mitigate the impacts of water use, if it identifies problems. The CEC's goal is to complete the entire certification process in 12 months, but public objections, incomplete application submittals, staff shortages, and limited budgets sometimes delay the process.

The CEC evaluates several sources of water data before certifying plant applicants' water use. These include:

- the developer's proposals;
- data from the Department of Water Resources' groundwater database on water availability and water quality;
- U.S. Geological Survey data on water availability through its streamflow and groundwater monitoring programs and any specific basin studies;
- the State Water Resources Control Board's information on surface and groundwater quality; and
- computer groundwater models that analyze the long-term yield of the basin.

With respect to water quality, the CEC coordinates the issuance of permits relating to water quality for new power plants, but the State Water Resources Control Board sets overall state policy. The Board operates under authority delegated to it by the U.S. Environmental Protection Agency to implement certain federal laws, including the Clean Water Act, as well as authority provided under state laws designed to protect water quality and ensure that the state's water is put to beneficial uses. Nine Regional Water Boards are delegated responsibility for implementing the statewide water quality control plans and policies, including setting discharge requirements for permits for the National Pollutant Discharge Elimination System Program and issuing the permits.

Table 6. Power Plants Implemented, Approved or Planned Since January 1, 2004, by Cooling Type

Category[a]	Number of plants	Dry cooled	Wet recirculating cooling system		
			Freshwater	Reclaimed water	Impaired groundwater
Operational Plant[b,c,d,g]	7	0	3	4	1
Approved by the CEC but not yet operational[c,e,g]	10	3	2	4	2
Currently under CEC review[f]	20	11	1	7	1
Total[d,e,g]	37	14	6	15	4

Source: GAO analysis of data from the California Energy Commission for plants sited, approved, or currently under review between January 1, 2004, and April 30, 2009.

[a] Excludes simple cycle gas plants with no steam cycle.
[b] Plants that started operating after 1/1/2004. These plants may have been approved by the CEC earlier.
[c] Includes one geothermal plant.
[d] One plant uses both recycled and impaired groundwater.
[e] Includes one hybrid plant that combines dry and wet cooling.
[f] Includes 7 solar thermal plants.
[g] Totals do not equal due to several plants using multiple water or cooling sources. See notes d and e.

Recent State Decisions and Current Proposals about Power Plant Water Use

Since 2004, most power plants the CEC has approved or is currently reviewing plan to use dry cooling or a wet recirculating system that uses an alternative water source, as shown in table 6. According to a state official we spoke with, no plants approved to be built in the last 25 years have used once-through cooling technology. Over the last 7 years, the CEC has also commissioned, or been involved in, substantial research into the use and possible effects of using alternative cooling technologies.

APPENDIX IV. REVIEW OF PROPOSALS TO USE WATER IN NEW POWER PLANTS IN GEORGIA

Background

In 2008, Georgia ranked 9th in population among states, with 9.7 million people, and had the 4th fastest growing population in the U.S. between the years 2000 and 2007. Georgia is historically water rich, receiving approximately 51 inches of precipitation annually, but recent droughts and growing population have prompted additional focus on water supply and management strategies. Georgia ranked 8th in total net electricity generation in 2007, accounting for approximately 3.5 percent of net electricity generation in the United States. Coal and nuclear power are the primary fuel sources for electricity in Georgia, with coal-fired power plants providing more than 60 percent of electricity output.

Georgia Water Law and Policy

Georgia is a regulated riparian state, meaning that the owners of land adjacent to a water body can choose when, where, and how to use the water. The use must be considered reasonable relative to a competing user, with the courts responsible for resolving disputes about reasonable use. Since the late 1970s, Georgia law has required any water user who withdraws more than an average of 100,000 gallons per day to obtain a withdrawal permit from the Georgia Environmental Protection Division.[54]

Georgia does not have a policy or guidance specifically addressing thermoelectric power plants' water use. However, in response to recent droughts and population growth, the state adopted its first statewide water management plan in 2008. State water regulators we spoke with said they expect the new state water plan to consider how future power generation siting decisions align with state water supplies.

Certification and Water Permitting for New Power Plants

Before power plant developers can begin construction, they may be required to obtain certification from the Georgia Public Service Commission and relevant permits from offices such as the Georgia Environmental Protection Division, as follows:

- *Georgia Public Service Commission.* Georgia Power Company, the state's investor-owned utility, is fully regulated by the Public Service Commission and must obtain a certificate of public convenience and necessity prior to constructing new power plants. Other power plant developers, including municipality- and cooperatively-owned power plants and others, are not subject to certification. Public Service Commission officials explained that during the certification process, they balance the need for the new plant and its costs, but they do not consider the impact a plant will have on Georgia's water supply. However, these officials explained that, in their capacity to ensure utilities charge just and reasonable rates, they could consider the economic impact of using an alternative water source or advanced cooling technology, should a plant propose to use one.
- *Georgia Environmental Protection Division.* Any entity seeking to use more than 100,000 gallons of water per day, including power plant developers, must obtain a permit from the Georgia Environmental Protection Division. The Division analyzes the proposed quantity of withdrawals and the water source and determines whether the withdrawal amounts and potential effects for downstream water users are acceptable. In some instances, the Division may place special conditions on power plants to ensure adequate water availability, such as requiring on-site reservoirs or groundwater withdrawals for water use during droughts. In making their decisions, the Georgia Environmental Protection Division reviews the plant's application and hydrologic data from a number of sources. Water withdrawal applications include many factors, in addition to withdrawal amounts and sources, such as water conservation and drought contingency plans; documentation of growth in water demand, location, and purpose of water withdrawn or diverted; and annual consumption estimates. Other data sources include their own and U.S. Geological Survey (USGS) groundwater data, USGS streamflow data, and existing water use permits. In some instances, the Environmental Protection Division may also use water withdrawal and water quality data collected by the U.S. Army Corps of Engineers if an applicant is downstream of federally-regulated waters. In addition to permitting water use, the Division is also responsible for issuing and enforcing all state permits involving water quality impacts. It is authorized by the Environmental Protection Agency to issue National Pollutant Discharge Elimination System permits that address discharge limits and reporting requirements.

Recent State Decisions about Power Plant Water Use

According to Division officials, the Division has never denied a water withdrawal permit to a power plant developer on the basis of insufficient water, which they attributed partly to the fact that the staff meets with applicants numerous times before they submit the application to identify and mitigate concerns about water availability. Moreover, they told us that thermoelectric power plant developers have submitted few applications for water withdrawal permits. For example, as shown in table 7, between January 1, 2004, and December 31, 2008, the Division received only 6 water withdrawal applications from thermoelectric power plant developers; of these, it approved 5. An official from the Public Service Commission was unaware of any regulated power plant developers proposing the use of advanced cooling technologies, such as dry cooling or hybrid cooling, over this time period.

Georgia Environmental Protection Division officials told us they do not advocate or refuse the use of particular cooling technologies. However, officials said they do not expect to receive applications for once-through cooling plants because federal environmental regulations make the permitting process difficult.

Table 7. Thermoelectric Power Plant Applications for Water Withdrawal Permits in Georgia Between January 2004 and December 2008

Category	Number of Plants	Once-through	Recirculating Groundwater (Freshwater)	Recirculating Surface water (Freshwater)	Recirculating Reclaimed water
Applied	6	0	4	2	1
Permitted[a]	5	0	3	1	1

Source: GAO analysis of data provided by the Georgia Environmental Protection Division.

Note: Totals do not equal due to one power plant developer submitting both a groundwater and surface water withdrawal application.

[a] As of August 12, 2009, one plant's application is still pending a decision by the Georgia Environmental Protection Division.

APPENDIX V. LIMITATIONS TO FEDERAL WATER USE DATA IDENTIFIED BY THOSE GAO CONTACTED

Data source	Limitation	Cause	Effect
EIA	*Advanced cooling technologies:* Data users cannot comprehensively identify plants making use of advanced cooling technologies, such as dry and hybrid cooling.	EIA forms are not designed to collect information on advanced cooling technologies.	Understanding of trends in the adoption of advanced cooling technologies cannot be systematically determined using only EIA data.
EIA	*Cooling system codes:* Codes used to classify plant cooling systems may be incomplete, lack explanation, overlap, or contain errors.	Cooling system codes are not defined in detail and plants may be uncertain about what cooling system code to use.	Inconsistent use of cooling tower codes could potentially make EIA data less valuable and lead to inaccurate or inconsistent data and analysis.
EIA	*Nuclear water data:* Water use data (withdrawal, consumption and discharge) and cooling information were discontinued for nuclear plants in 2002.	EIA discontinued reporting nuclear water use data and cooling system information due to priorities stemming from budget limitations.	Data users must use noncurrent data or seek out an alternate source. If this limitation persists, water data will not be available for any new nuclear plants constructed.

(Continued)

EIA and USGS	*Alternative water sources:* It is not possible to comprehensively identify power plants using alternative water sources.	EIA forms are not design-ned to collect information on alternative water sources. According to USGS, budget constraints have limited the amount of water use information the agency can provide.	Understanding trends in power plant adoption of alternative water sources is limited.
EIA and USGS	*Frequency:* EIA reports data on annual water use, rather than data on water use over shorter time periods, such as monthly. USGS reports 5-year data.	EIA's form 767, used to collect cooling system and water data, was developed and revised in the 1980s, and EIA officials we spoke with were not aware of why an annual time period was originally chosen. According to USGS, budget constraints have limited the amount of water use information the agency can provide.	Seasonal trends in water use by power plants are not evident from annual EIA or 5-year USGS data.
EIA and USGS	*Quality:* Reporting of some EIA data elements may be inaccurate or inconsistent. USGS data are compiled from many different data sources, and the accuracy and methodology of these sources may vary. Furthermore, USGS state offices have different methods for developing water use estimates, potentially contributing to data inconsistency.	Respondents may use different methods to measure or estimate data and instructions may be limited or unclear. Respondents may make mistakes or have non-technical staff fill out surveys, since EIA's form for collecting this data does not require technical staff to complete the survey. According to USGS, budget constraints in its water use program kept the agency from implementing improvements it would like to make to its quality control of water use data.	Inaccurate and incon-sistent data are more challenging to analyze and less relevant for policymakers, water experts and the public seeking to understand water use patterns.
USGS	*Consumption:* USGS discontinued reporting of thermoelectric power plant and other water consumption data.	According to USGS, budget constraints have caused the agency to make cuts in data reporting.	Understanding of trends in power plant water consumption compared to other industries is limited. Analysis to compare thermoelectric power plant withdra-wals to consumption is more complicated.
USGS	*Hydrologic code:* USGS discontinued reporting thermoelectric power plant and other water use by hydrologic code. It now only reports data by county.	According to USGS, budget constraints have caused the agency to make cuts in data reporting.	According to some data users, not having data by hydrologic code complicates water analysis, which is often performed by watershed rather than county.
USGS	*Timeliness:* Data are reported many years late. For example, data on 2005 water use have not yet been made available to the public.	According to USGS, budget constraints have led to limited staff availability for water use data collection and analysis, resulting in reporting delays.	Data are outdated and may be less relevant for analysis.

Source: GAO analysis of comments gathered during interviews with water and electricity experts, environmental groups, and federal agencies.

APPENDIX VI. COMMENTS FROM THE DEPARTMENT OF THE INTERIOR

United States Department of the Interior

OFFICE OF THE SECRETARY
Washington, D.C. 20240

SEP 29 2009

Ms. Anu Mittal
Director, Natural Resources and Environment
U.S. Government Accountability Office
441 G Street, N.W.
Washington, D.C. 20548

Dear Ms. Mittal:

Thank you for providing the Department of the Interior (DOI) the opportunity to review and comment on the draft Government Accountability Office (GAO) Report entitled, *"ELECTRICITY AND WATER: Improvements to Federal Water Use Data Would Increase Understanding of Trends in Power Plant Water Use" (GAO-09-912).*

The DOI agrees with the recommendations made by the GAO. The USGS works in cooperation with local, State, and Federal agencies to compile and disseminate data on the Nation's water use. Enhancement of water-use information is a key element of the Subtitle F-Secure Water of the Omnibus Public Lands Management Act of 2009 (P.L. 111-11) and is a high priority component of the Water Census of the United States, one of six strategic science directions for the USGS. As information becomes available from the Energy Information Administration (EIA), the USGS will expand efforts to disseminate data on the use of alternative water sources by thermoelectric power plants. The USGS views water consumption data at thermoelectric plants as an important component of the Water Census and will reinstate its collection as future resources allow. The USGS will coordinate with EIA to establish a process to identify and implement steps to improve water-use data collection and dissemination by the two agencies.

We hope these comments will assist you in preparing the final report. If you have any questions, or need additional information, please contact Dr. Matt Larsen (703) 648-5215 or Mr. William Cunningham at (703) 648-5005.

Sincerely,

Anne J. Castle
Assistant Secretary for
Water and Science

End Notes

[1] The Environmental Protection Agency announced in a September 15, 2009, press release its plans to revise existing standards for water discharges from coal-fired power plants.

[2] Pub. L. No. 111-11, § 9508 (2009).

[3] S. 531, 111th Cong. § 2 (2009).

[4] H.R. 3598, 111th Cong. (2009).

[5] We provided preliminary information from our work on two of these reports—biofuels and water use and thermoelectric power plants and water use—in a testimony before the Subcommittee on Energy and Environment in July 2009. GAO, *Energy and Water: Preliminary Observations on the Links between Water and Biofuels and Electricity Production.* (Washington, D.C.: July 9, 2009). GAO-09-862T.

[6] Studies we reviewed indicated a range of temperature increases for water discharged from once-through cooling systems. EPA officials we spoke with told us that once-through cooling plants often discharge cooling water between 10 and 20 degrees Fahrenheit warmer than it was when it was withdrawn, but they explained that there are examples of plants above and below this range, as well.

[7] Another method of dry cooling, referred to as indirect dry cooling, uses a closed-loop of cooling water to condense the steam exiting the turbine—similar to recirculating systems. However, instead of dissipating the cooling water's heat through evaporation, a dry cooling tower is used to transfer the heat from the cooling water to the ambient air.

[8] Some experts we spoke with and documents we reviewed described two other types of hybrid cooling technology designs. One version is designed to minimize plumes released from wet recirculating systems with cooling towers; although, according to one expert, this version has very little effect on the plant's water consumption. The other consists of various system configurations designed to improve the efficiency of dry cooling by either spraying water on the air-cooled condenser directly or using water to lower the temperature of inlet air entering the air-cooled condenser.

[9] Department of Energy, National Energy Technology Laboratory, *Estimating Freshwater Needs to Meet Future Thermoelectric Generation Requirements*. 2008. This report did not include statistics regarding the use of hybrid systems.

[10] Electric Power Research Institute, *Water Use for Electric Power Generation*, (Palo Alto, CA, 2008). 1014026.

[11] Argonne National Laboratory, *Use of Reclaimed Water for Power Plant Cooling*, (Argonne, IL., 2007).

[12] Energy is also needed in wet recirculating systems with fan-forced cooling towers, as well as to operate water pumps in both once-through and wet recirculating systems with cooling towers. Wet recirculating systems with cooling towers can also be constructed with a type of cooling tower that relies on a chimney effect, rather than fans, to naturally produce airflow. These natural draft cooling towers are large concrete structures that are significantly more expensive to build than cooling towers with fans, although they would eliminate the energy costs associated with fan operation.

[13] Environmental Protection Agency, *Technical Development Document for the Final Regulations Addressing Cooling Water Intake Structures for New Facilities*, (Washington, D.C., Nov. 2001). These figures were higher for a full steam fossil fueled plant, such as a coal plant. Representatives from EPA explained that energy penalty and cost comparisons between dry cooled systems and wet recirculating systems with cooling towers may have changed since EPA's 2001 report was issued. The agency is in the process of updating its estimates of energy penalties and cooling system costs.

[14] Burns, John M. and Wayne Micheletti, *Emerging Issues and Needs in Power Plant Cooling Systems*. (Presented at DOE's Workshop on Electric Utilities and Water: Emerging Issues and Needs, Pittsburgh, PA, July 23-24, 2002).

[15] Plants with once-through systems and wet recirculating systems with cooling towers also face efficiency losses as water and wet-bulb temperatures rise. As noted, dry cooled plants tend to be less efficient than plants with both of these wet cooling systems, but the efficiency of dry cooled plants will approach that of wet cooled plants at certain times of the year and in certain climatic conditions. For example, according to experts we spoke with, there will be a smaller difference in efficiency between a plant with a wet recirculating system with cooling towers and a dry cooled plant in cool, humid climates.

[16] We include examples from these studies to provide context about the magnitude of estimated energy penalties. We have not validated the methodology or results of these studies. Estimates are subject to study assumptions and methodology, and actual energy penalties depend highly on plant design, location, and decisions made by plant developers about how to optimize total plant costs.

[17] Environmental Protection Agency, *Technical Development Document for the Final Regulations Addressing Cooling Water Intake Structures for New Facilities*, (Washington, D.C., Nov. 2001). EPA estimated energy penalties at peak summer conditions when plants operate at 100 percent capacity to be higher. For example, the study estimates national average energy penalties at peak summer conditions (100 percent capacity) to result in 2.4 percent lower output for combined cycle plants with dry cooling systems compared to those with wet recirculated systems with a cooling tower. EPA estimated national average energy penalties at peak summer conditions (100 percent capacity) to result in 8.4 percent lower output for full steam fossil fueled plants, such as coal plants, with dry cooling systems, compared to those with wet recirculated systems with a cooling tower. Representatives from EPA explained that energy penalty and cost comparisons between dry cooled systems and wet recirculating systems with cooling towers may have changed since EPA's 2001 report was issued. The agency is in the process of updating its estimates of energy penalties and cooling system costs.

[18] Department of Energy, Office of Fossil Energy, National Energy Technology Laboratory and Argonne National Laboratory, *Energy Penalty Analysis of Possible Cooling Water Intake Structure Requirements on Existing Coal-Fired Power Plants*. (2002). These estimates refer to a dry cooling tower with a 20 degree Fahrenheit approach, the difference between the air temperature and the temperature of cold water discharged from the condenser. Energy penalty estimates for a dry tower with a 40 degree Fahrenheit approach were higher. The 1 percent hottest day estimate is for plants with a range of 15 degrees Fahrenheit, where the range refers to the difference between the temperature of the water entering and leaving the condenser. This study focused on

existing plants retrofitted with indirect dry cooled systems, which are considered less efficient than direct dry systems. Experts we spoke with told us energy penalties are higher in retrofitted plants than when a dry cooled system is designed according to the unique specifications of a newly built plant because indirect dry cooling systems are more likely to be used; plant components, like the turbine, have not been designed to work most effectively with a dry cooled system; and because of size constraints placed on the dry cooled system.

[19] Hot day performance is estimated to be the 1 percent highest dry bulb temperature and the corresponding wet bulb temperature for that condition.

[20] We include examples of cost estimates from selected studies and expert interviews in this section to provide context about the magnitude of estimated capital and operating costs of dry cooling systems compared to wet cooling systems. We have not validated the methodology or results of these estimates. Estimates are subject to each study's assumptions and methodology, and actual costs depend highly on plant design, locational factors such as water costs, and decisions made by plant developers about how to optimize total costs. Furthermore, it should be noted that cooling system costs are but one component of total plant costs.

[21] Electric Power Research Institute, *Comparison of Alternate Cooling Technologies for U.S. Power Plants. Economic, Environmental, and Other Trade-offs,* (Palo Alto, CA., 2004). 1005358. Similarly, capital costs for a dry cooled system on theoretical 350 megawatt coal plants ranged between $43 and $47 million for 5 climatic locations—3.2 to 3.6 times that of a wet recirculating system with cooling tower.

[22] Electric Power Research Institute, *Comparison of Alternate Cooling Technologies for U.S. Power Plants. Economic, Environmental, and Other Trade-offs,* (Palo Alto, CA., 2004). 1005358.

[23] Electric Power Research Institute, *Water Use for Electric Power Generation.*

[24] These users are issued a Certificate of Use, indicating the use has been registered with the State of Alabama. All Certificate of Use holders are required to annually report their water usage to the Alabama Office of Water Resources.

[25] If a prospective water user is unable to acquire a new water right, he or she may choose to purchase or lease an existing water right.

[26] In the two cases we identified, an official from the Public Utilities Commission of Nevada told us the power plants in question were never built. He also noted that as many as six power plants have been sited in Nevada with dry cooling due to lack of available water.

[27] Examples of restrictions include 1) to not maliciously injure a neighbor, 2) to not willfully waste water, 3) to not drill a well slanting under a neighbor's property or 4) to assume liability for damages for negligent pumping that causes subsidence of a neighboring land.

[28] California also has local districts, known as Adjudicated Groundwater Basins, that may impose similar requirements.

[29] In 1975, the State Water Resources Control Board established a policy that inland freshwater should be considered the water type of last resort for power plants and encouraged utilities to study the feasibility of effluent from sewage treatment plants for power plant cooling. The policy states the use of fresh inland waters for power plant cooling will only be approved when it is demonstrated that the use of other water supply sources or other methods of cooling would be environmentally undesirable or economically unsound.

[30] The California Energy Commission reiterated the 1975 policy in the December 2003 Integrated Energy Policy Report that, consistent with that 1975 State Water Resources Control Board policy, it would only approve the use of freshwater where alternative cooling technologies were shown to be "environmentally undesirable" or "economically unsound."

[31] One of these power plants uses a hybrid cooling system and is counted as having a water source and as using dry cooling.

[32] Simple cycle natural gas plants are excluded from these statistics since they do not have a steam cycle and, therefore, do not need water for cooling.

[33] There are variations for different plants in the number of cycles required and exemptions for the first full year of operation.

[34] The most significant loss of water in a wet recirculating system with cooling towers is through evaporation from cooling towers. However, studies conducted by the Electric Power Research Institute indicate that increasing the cycles of concentration can result in water savings, though with diminishing returns after a certain number of cycles.

[35] Commission officials noted that their review may indirectly affect a power plant's water use since consideration of cooling systems can be one component in their consideration of a power plant's feasibility, reliability and cost. In general, the Commission will favor the least-cost cooling option that ensures electric reliability and defers to state water agencies to address issues related to a plant's potential impact on water quality and quantity. However, officials also explained there may be circumstances where cooling or water issues are raised in a public hearing that may need to be considered by the Commission.

[36] The Line Siting Committee makes a recommendation to the Arizona Corporation Commission about whether to issue a Certificate of Environmental Compatibility. The Arizona Corporation Commission is responsible for the final approval, modification, or denial of the certificate.

[37] Power plants planning to use surface water must have surface water rights approved by the State Water Resources Control Board. Board officials told us that recent power plant applications for surface water rights were rare. According to an official at the California Energy Commission, power plants planning to use surface water often obtain their supply through a retail water agency, rather than obtaining surface water rights directly.

[38] Unlike federal data on water availability, federal data on water use developed by USGS and EIA is not routinely relied upon by representatives from most of the state water regulators we spoke with, who evaluate applications for water use permits and water rights for new power plants. Some said they, instead, used data their offices had developed internally, including water use data reported to them by water permit and rights holders.

[39] The Arizona Department of Water Resources collects water use data from water users in Active Management Areas, which are statutorily designated areas of constrained water supply. However, according to one official, the Department does not generally have the ability to collect these data outside of Active Management Areas and Irrigation Non-expansion Areas. Instead, the Department has entered into a cooperative agreement with USGS to collect these data.

[40] Electric Power Research Institute, *Comparison of Alternate Cooling Technologies for California Power Plants: Economic, Environmental and Other Tradeoffs,* (Palo Alto, CA., 2002).

[41] EIA officials noted that the agency collects environmental information from all U.S. plants with an existing or planned organic-fueled or combustible renewable stream-electric unit with a generator nameplate rating of 10 megawatts or larger. Form 767 instructions require cooling system and water information to be reported by plants with a nameplate capacity of 100 megawatts or greater.

[42] EIA reports water consumption data for plants 100 megawatts in size or larger, but has not published aggregated data in such a way that allows them to be readily used to identify overall trends in thermoelectric power plant water consumption compared to withdrawal. However, these and other environmental data collected by EIA from 1996 to 2005 for individual plants are available on EIA's Web site and can be assessed by all users at http://www.eia.doe.gov/cneaf/electricity/page/eia767.html.

[43] Warm water discharged back into a water body from a once-through system may increase evaporation—water consumption—from the receiving water body. One expert we spoke with suggested that including this indirect form of water consumption in plant estimates would improve the federal data.

[44] National Research Council, *Estimating Water Use in the United States* (Washington, D.C., 2002).

[45] According to an Arizona Department of Water Resources official, it issues a Certificate of Water Right once the water is put to beneficial use. Several areas of decreed rights exist, for example, Globe Equity Decree on the Upper Gila River.

[46] According to Arizona Department of Water Resources officials, options for obtaining groundwater rights include the following: (1) an existing Irrigation Grandfathered Groundwater Right that can be legally retired to a Type 1 Non-Irrigation Grandfathered Groundwater Right (A.R.S. § 45-469); (2) an existing Type 1 Non-Irrigation Grandfathered Groundwater Right (A.R.S. §§ 45-470, 45-472, 45-473, 45-542)); (3) a Type 2 Non-Irrigation Grandfathered Groundwater Right, which can be purchased or leased from another owner within the same Active Management Area (A.R.S. § 45-471); or (4) a General Industrial Use Permit, a permit to pump groundwater from a point outside of the exterior boundaries of the service area of a city, town, or private water company for non-irrigation purposes (A.R.S. § 45-515). Inside the Harquahala Irrigation Non-Expansion Area, there are some limitations to pumping groundwater for industrial uses, pursuant to A.R.S. § 45-440.

[47] There are variations for different plants in the number of cycles required and exemptions for the first full year of operation.

[48] Plants smaller that 100 megawatts do not need state siting approval. However, they must still comply with any and all local ordinances or state ordinances such as zoning, water quality, air quality, etc.

[49] The Committee is chaired by a representative from the Office of the Arizona Attorney General. Other agencies represented include the Department of Environmental Quality, Department of Water Resources, the Office of Energy in the Department of Commerce, and the Arizona Corporation Commission.

[50] Due to declining electricity prices, some of the approved plants were never constructed and others were sold to new owners.

[51] In some areas of California, groundwater is managed locally through Adjudicated Groundwater Basins that can regulate the amount of groundwater extracted.

[52] California's local air pollution control and air quality management districts have the authority to issue construction permits for the operation of power plants with less than 50 megawatts of generating capacity.

[53] Though not common, if a power plant developer plans to make use of surface water in California, it may be required to apply for a water right from the State Water Resources Control Board. In evaluating the permit application, the State Water Resources Control Board would conduct its own analysis using a combination of state and federal data sources.

[54] Any entity that withdraws more than 100,000 gallons a day (monthly average) of surface water or 100,000 gallons a day (daily average) of groundwater requires a water permit from the Division.

In: Exploring the Energy-Water Nexus
Editor: Peter D. Wright
ISBN: 978-1-61209-791-6
© 2011 Nova Science Publishers, Inc.

Chapter 3

MANY UNCERTAINTIES REMAIN ABOUT NATIONAL AND REGIONAL EFFECTS OF INCREASED BIOFUEL PRODUCTION ON WATER RESOURCES

United States Government Accountability Office

WHY GAO DID THIS STUDY

In response to concerns about the nation's energy dependence on imported oil, climate change, and other issues, the federal government has encouraged the use of biofuels. Water plays a crucial role in all stages of biofuel production—from cultivation of feedstock through its conversion into biofuel. As demand for water from various sectors increases and places additional stress on already constrained supplies, the effects of expanded biofuel production may need to be considered.

To understand these potential effects, GAO was asked to examine (1) the known water resource effects of biofuel production in the United States; (2) agricultural conservation practices and technological innovations that could address these effects and any barriers to their adoption; and (3) key research needs regarding the effects of water resources on biofuel production. To address these issues, GAO reviewed scientific studies, interviewed experts and federal and state officials, and selected five states to study their programs and plans related to biofuel production.

GAO is not making any recommendations in this report. A draft of this report was provided to the Departments of Agriculture (USDA), Energy (DOE), and the Interior (DOI); and the Environmental Protection Agency (EPA). USDA, DOE, and DOI concurred with the report and, in addition to EPA, provided technical comments, which were incorporated as appropriate.

WHAT GAO FOUND

The extent to which increased biofuels production will affect the nation's water resources depends on the type of feedstock selected and how and where it is grown. For example, to the extent that this increase is met from the cultivation of conventional feedstocks, such as corn, it could have greater water resource impacts than if the increase is met by next generation feedstocks, such as perennial grasses and woody biomass, according to experts and officials. This is because corn is a relatively resource-intensive crop, and in certain parts of the country requires considerable irrigated water as well as fertilizer and pesticide application. However, experts and officials noted that next generation feedstocks have not yet been grown on a commercial scale and therefore their actual effects on water resources are not fully known at this time. Water is also used in the process of converting feedstocks to biofuels, and while the efficiency of biorefineries producing corn ethanol has increased over time, the amount of water required for converting next generation feedstocks into biofuels is still not well known. Finally, experts generally agree that it will be important to take into account the regional variability of water resources when choosing which feedstocks to grow and how and where to expand their production in the United States.

The use of certain agricultural practices, alternative water sources, and technological innovations can mitigate the effects of biofuels production on water resources, but there are some barriers to their widespread adoption. According to experts and officials, agricultural conservation practices can reduce water use and nutrient runoff, but they are often costly to implement. Similarly, alternative water sources, such as brackish water, may be viable for some aspects of the biofuel conversion process and can help reduce biorefineries' reliance on freshwater. However, the high cost of retrofitting plants to use these water sources may be a barrier, according to experts and officials. Finally, innovations—such as dry cooling systems and thermochemical processes—have the potential to reduce the amount of water used by biorefineries, but many of these innovations are currently not economically feasible or remain untested at the commercial scale.

Many of the experts GAO spoke with identified several areas where additional research is needed. These needs fall into two broad areas: (1) feedstock cultivation and biofuel conversion and (2) data on water resources. For example, some experts noted the need for further research into improved crop varieties, which could help reduce water and fertilizer needs. In addition, several experts identified research that would aid in developing next generation feedstocks. For example, several experts said research is needed on how to increase cultivation of algae for biofuel to a commercial scale and how to control for potential water quality problems. In addition, several experts said research is needed on how to optimize conversion technologies to help ensure water efficiency. Finally, some experts said that better data on water resources in local aquifers and surface water bodies would aid in decisions about where to cultivate feedstocks and locate biorefineries.

ABBREVIATIONS

CRP	Conservation Reserve Program
NPDES	National Pollutant Discharge Elimination System

DOE	Department of Energy
EIA	Energy Information Administration
EISA	Energy Independence and Security Act of 2007
EPA	Environmental Protection Agency
RFS	Renewable Fuel Standard
USDA	U.S. Department of Agriculture
USGS	U.S. Geological Survey
UST	underground storage tank

November 30, 2009

The Honorable Bart Gordon
Chairman
Committee on Science and Technology
House of Representatives

Dear Mr. Chairman:

In recent years, the federal government has increasingly encouraged the use of biofuels and other alternatives to petroleum in response to concerns over U.S. dependence on imported oil, climate change, and other issues. The United States is the largest user of petroleum in the world, consuming 19.4 million barrels per day in 2008, over half of which is imported. Biofuels, such as ethanol and biodiesel, can be produced domestically and are derived from renewable sources, such as corn, sugar cane, and soybeans. The Energy Independence and Security Act of 2007 (EISA) expanded the Renewable Fuel Standard (RFS) by requiring that U.S. transportation fuel contain 9 billion gallons of renewable fuels in 2008 and increasing this amount annually to 36 billion gallons in 2022.[1] Currently, the vast majority of domestic biofuel production is ethanol derived from corn starch, which EISA defines as a "conventional" feedstock. However, in 2022, the RFS's 36-billion-gallon total requires that at least 16 billion gallons be derived from "cellulosic" materials, such as stalks, stems, branches, and leaves. These cellulosic materials, along with newer feedstocks, such as algae, are often referred to as "next generation" feedstocks, and the fuels produced from them are often referred to as "advanced" biofuels.[2]

Although freshwater flows abundantly in many of the nation's lakes, rivers, and streams, water is a dwindling resource in many parts of the country and is not always available when and where it is needed or in the amount desired because of competing demands on water supplies, climatic changes contributing to drought conditions in parts of the country, and population growth. Foremost among these competing demands is irrigation, which accounts for 40 percent of the nation's freshwater withdrawals.[3] Water is crucial to many stages of the biofuel life cycle and is needed for the growth of the feedstock as well as for fermentation, distillation, and cooling during the process of converting the feedstock into biofuel. As biofuel production increases, questions have emerged about the effects that increased production could have on the nation's water resources.

To understand the potential effects of increased biofuel production on water resources, you asked us to describe (1) the known water resource effects of increased biofuel production in the United States; (2) the agricultural conservation practices and technological innovations that exist or are being developed to address these effects, and any barriers that may prevent the adoption of these practices and technologies; and (3) key research needs regarding the effects of biofuel production on water resources.

To address all of these objectives, we conducted a systematic analysis of relevant articles from scientific journals and key federal and state government publications. In addition, in consultation with the National Academy of Sciences, we identified and interviewed recognized experts who have published peer-reviewed research analyzing the water supply requirements of one or more biofuel feedstocks and the implications of increased biofuel production on water resources. These experts included research scientists in such fields as environmental science, agronomy, soil science, hydrogeology, ecology, and engineering. Furthermore, we studied five states in greater depth—Georgia, Iowa, Minnesota, Nebraska, and Texas—to gain an understanding of the programs and plans they have or are developing to address increased biofuel production. We selected these states based on several criteria, including ethanol and biodiesel production, feedstock cultivation type, reliance on irrigation, geographic diversity, and varying approaches to water resource management and law. For each of the states, we analyzed documentation from and conducted interviews with a wide range of stakeholders to gain the views of diverse organizations covering all stages of biofuel production. These groups included relevant state agencies, including those responsible for oversight of agriculture, environmental quality, and water and soil resources; federal agency officials with responsibility for a particular state or region, such as officials from the U.S. Geological Survey (USGS), the U.S. Department of Agriculture's (USDA) Natural Resources Conservation Service, and the Environmental Protection Agency (EPA); university researchers; industry representatives; and relevant nongovernmental organizations, such as environmental groups, state-level corn growers' associations, and ethanol producer associations.

We also interviewed senior officials, scientists, economists, researchers, and other federal officials from USDA, the Departments of Defense and Energy (DOE), EPA, the National Aeronautics and Space Administration, the Department of Commerce's National Oceanic and Atmospheric Administration, the National Science Foundation, and USGS about effects on water supply and water quality during biofuel production. We also interviewed representatives of nongovernmental organizations, such as the Renewable Fuels Association, the Biotechnology Industry Organization, the Pacific Institute, and the Fertilizer Institute. A more detailed description of our scope and methodology is presented in appendix I. We conducted our work from January 2009 to November 2009 in accordance with all sections of GAO's Quality Assurance Framework that are relevant to our objectives. The framework requires that we plan and perform the engagement to obtain sufficient and appropriate evidence to meet our stated objectives and to discuss any limitations in our work. We believe that the information and data obtained, and the analysis conducted, provide a reasonable basis for any findings and conclusions in this product.

BACKGROUND

Biofuels, such as ethanol and biodiesel, are an alternative to petroleum-based transportation fuels and are produced in the United States from a variety of renewable sources such as corn, sugar cane, and soybeans. Ethanol, the most common U.S. biofuel, is mainly used as a gasoline additive in blends of about 10 percent ethanol and 90 percent gasoline, known as E10, which is available in most states. A relatively small volume is also blended at a higher level called E85—a blend of 85 percent ethanol and 15 percent gasoline—which can only be used in specially designed vehicles, known as flexible fuel vehicles. Biodiesel is a renewable alternative fuel produced from a range of plant oils, animal fats, and recycled cooking oils. Pure biodiesel or biodiesel blended with petroleum diesel—generally in a blend of 20 percent biodiesel and 80 percent diesel—can be used to fuel diesel vehicles.

The federal government has promoted biofuels as an alternative to petroleum-based fuels since the 1970s, and production of ethanol from corn starch reached 9 billion gallons in 2008. The Energy Policy Act of 2005 originally created an RFS that generally required U.S. transportation fuel to contain 4 billion gallons of renewable fuels in 2006 and 7.5 billion gallons in 2012.[4] EISA expanded the RFS by requiring that U.S. transportation fuel contain 9 billion gallons of renewable fuels in 2008 and increasing this amount annually to 36 billion gallons in 2022.[5] Moreover, the 36-billion-gallon total must include at least 21 billion gallons of advanced biofuels, defined as renewable fuels other than ethanol derived from corn starch that meet certain criteria; only 15 billion of the 36 billion gallons of renewable fuels can come from conventional biofuels. In addition, at least 16 billion gallons of the 21-billion-gallon advanced biofuels requirement must be made from cellulosic feedstocks, such as perennial grasses, crop residue, and woody biomass. Unlike corn starch, most of the energy in plant and tree biomass is locked away in complex cellulose and hemicellulose molecules, and technologies to produce biofuels economically from this type of feedstock are still being developed. Some cellulosic biorefineries are piloting the use of biochemical processes, in which microbes and enzymes break down these complex plant molecules to produce ethanol, while others are piloting the use of thermochemical processes, which use heat and chemical catalysts to convert plant material into a liquid that more closely resembles petroleum.

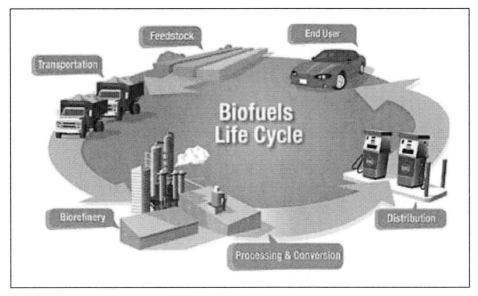

Source: DOE.

Figure 1. Biofuels Life Cycle.

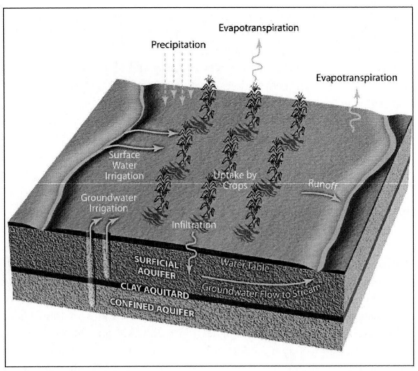

Source: © 2008 International Mapping.

Figure 2. Agricultural Water Cycle.

There are a number of steps in the biofuels life cycle, from cultivation of the feedstock through distribution to the end user at the fuel pump (see figure 1).

Water plays a critical role in many aspects of this life cycle. On the cultivation side, water is needed to grow the feedstock. Crops can be either rainfed, with all water requirements provided by natural precipitation and soil moisture, or irrigated, with at least some portion of water requirements met through applied water from surface or groundwater sources. Figure 2 shows the various water inputs (sources of water) and outputs (water losses) that are part of the agricultural water cycle.

Water is also important for conversion of feedstocks into biofuels. In particular, water is used for heating and cooling as well as for processing. For example, during the processing of corn-based ethanol, corn is converted to ethanol through fermentation using one of two standard processes, dry milling or wet milling. The main difference is the initial treatment of the corn kernel. In the dry-mill process, the kernel is first ground into flour meal and processed without separating the components of the corn kernel. The meal is then slurried with water to form a mash, and enzymes are added to convert the starch in the mash to a fermentable sugar. The sugar is then fermented and distilled to produce ethanol. In the wet-mill process, the corn kernel is steeped in a mixture of water and sulfurous acid that helps separate the kernel into starch, germ, and fiber components. The starch that remains after this separation can then be fermented and distilled into fuel ethanol. Traditional dry-mill ethanol plants cost less to construct and operate than wet-mill plants, but yield fewer marketable co-products. Dry-mill plants produce distiller's grains (that can be used as cattle feed) and carbon dioxide (that can be used to carbonate soft drinks) as co-products, while wet-mill plants produce many more co-products, including corn oil, carbon dioxide, corn gluten meal, and corn gluten feed. The majority of ethanol biorefineries in the United States are dry-mill facilities. Figure 3 depicts the conversion process for a typical dry-mill biorefinery.

EACH STAGE OF BIOFUEL PRODUCTION AFFECTS WATER RESOURCES, BUT THE EXTENT DEPENDS ON THE FEEDSTOCK AND REGION

The extent to which increased biofuel production will affect the nation's water resources will depend on which feedstocks are selected for production and which areas of the country they are produced in. Specifically, increases in corn cultivation in areas that are highly dependent on irrigated water could have greater impacts on water availability than if the corn is cultivated in areas that primarily produce rainfed crops. In addition, most experts believe that greater corn production, regardless of where it is produced, may cause greater impairments to water quality than other feedstocks, because corn production generally relies on greater chemical inputs and the related chemical runoff will impact water bodies. In contrast, many experts expect next generation feedstocks to require less water and provide some water quality benefits, but even with these feedstocks the effects on water resources will largely depend on which feedstock is selected, and where and how these feedstocks are grown. Similarly, the conversion of feedstocks into biofuels may also affect water supply and water quality, but these effects also vary by feedstock chosen and type of biofuel produced. Many experts agree that as the agriculture and biofuel production industries make decisions about which feedstocks to grow and where to locate or expand conversion facilities, it will be important for them to consider regional differences and potential impacts on water resources.

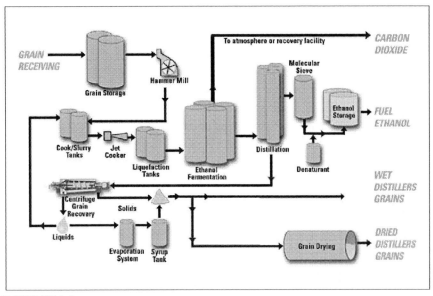

Source: © 2007 ICM, Inc.

Figure 3. Diagram of Conversion Process for a Typical Corn-Based Ethanol Biorefinery.

Table 1. Average Water Consumed in Corn Ethanol Production in Primary Producing Regions in the United States, in Gallons of Water/Gallon of Ethanol Produced

Type of water consumed	USDA Region 5 (Iowa, Indiana, Illinois, Ohio, Missouri)	USDA Region 6 (Minnesota, Wisconsin, Michigan)	USDA Region 7 (North Dakota, South Dakota, Nebraska, Kansas)
Cultivation			
Corn irrigation, groundwater	6.7	10.7	281.2
Corn irrigation, surface water	0.4	3.2	39.4
Total irrigated water	**7.1**	**13.9**	**320.6**
Conversion - Corn ethanol	3.0	3.0	3.0
Total water consumption	**10.0**	**16.8**	**323.6**

Source: Center for Transportation Research, Energy Systems Division, Argonne National Laboratory, "Consumptive Water Use in the Production of Ethanol and Petroleum Gasoline," Center for Transportation Research, Energy Systems Division, Argonne Laboratory, January 2009

Note: The numbers may not add up due to rounding. The Argonne National Laboratory study estimated the water consumed in corn ethanol production in each of the major ethanol producing regions considering water consumed in both corn cultivation and conversion processing steps. Estimates were based on average consumption of 3.0 gallons of water per gallon of corn ethanol produced in a corn dry mill, average consumptive use of irrigation water for corn in major corn producing regions, and dry-mill yield of 2.7 gallons of ethanol per bushel. In evaluating corn cultivation, the water consumed is based on total amount of irrigation water used for corn production and total corn production for each region, and does not include precipitation. In addition, the calculation assumes that 30 percent of water recharges local surface and groundwater, and the remaining 70 percent of the water is consumed by evapotranspiration (water lost through evaporation from the soil and plants) and other factors.

Water Supply and Water Quality Effects of Increased Corn Cultivation

Many experts and officials told us that corn cultivation requires substantial quantities of water, although the amount used depends on where the crop is grown and how much irrigation water is used. The primary corn production regions are in the upper and lower Midwest and include 12 states classified as USDA farm production Regions 5, 6, and 7. Together, these regions accounted for 89 percent of corn production in 2007 and 2008, and 95 percent of ethanol production in the United States in 2007. Corn cultivation in these three regions averages anywhere from 7 to 321 gallons of irrigation water for every gallon of ethanol produced, as shown in table 1.[6] However, the impact of corn cultivation on water supplies in these regions varies considerably. For example, in USDA Region 7, which comprises North Dakota, South Dakota, Kansas, and Nebraska, the production of one bushel of corn consumes an average of 865 gallons of freshwater from irrigation. In contrast, in USDA Regions 5 and 6, which comprise Iowa, Illinois, Indiana, Ohio, Missouri, Minnesota, Wisconsin, and Michigan, corn is mostly rainfed and only requires on average 19 to 38 gallons of supplemental irrigation water per bushel.[7]

The effects of increased corn production for ethanol on water supplies are likely to be greatest in water-constrained regions of the United States where corn is grown using irrigation. For example, some of the largest increases in corn acres (1.1 million acres) are projected to occur in the Northern Plains region, which is already a water constrained region. Parts of this region draw heavily from the Ogallala Aquifer, where water withdrawals are already greater than the natural recharge rate from precipitation. A 2009 USGS report found water levels in the aquifer had dropped more than 150 feet in parts of southwest Kansas and the Texas Panhandle, where crop irrigation is intense and recharge to the aquifer is minimal.[8] In 2000, about 97 percent of the water withdrawn from the aquifer was used for irrigation, according to USGS.[9]

Many officials told us that an increase in corn cultivation using current agricultural practices will also impair water quality as a result of the runoff of fertilizer into lakes and streams. This will happen because corn requires high applications of fertilizers relative to soybeans and other potential biofuel feedstocks, such as perennial grasses.[10] For example, in Iowa, the expansion of biofuel production has already led to an increasing amount of land dedicated to corn and other row crops, resulting in surface water impacts, including nutrient runoff and increased bacteria counts as well as leaching of nitrogen and phosphorus into groundwater, according to a state official. Fertilizer runoff containing nitrogen and phosphorus can lead to overenrichment and excessive growth of algae in surface waters. In some waters, such enrichment has resulted in harmful algal blooms, decreased water clarity, and reduced oxygen in the water, which impair aquatic life.[11] In marine waters, this excessive algal growth has created "dead zones," which cannot support fish or any other organism that needs oxygen to survive.[12] The number of reported dead zones around the world has increased since the 1960s to more than 400.[13] Many of them are along the Gulf of Mexico and the Atlantic Coast, areas that receive drainage from agricultural and urban landscapes, including a large portion of the Corn Belt, where many of the existing and planned ethanol production facilities are located. A 2007 USGS model estimated that 52 percent of the nitrogen and 25 percent of the phosphorus entering the Gulf system are from corn and soybean cultivation in the Mississippi River basin.[14]

Increased corn production will also increase the use of pesticides—including insecticides and herbicides—which also have the potential to affect surface water and groundwater quality. For example, a 10-year nationwide study by USGS detected pesticides in 97 percent of streams in agricultural and urban watersheds.[15] As would be expected, the highest concentrations of pesticides have been found in those areas that have the highest use. For instance, application rates of atrazine, a commonly used pesticide for corn production, are highest in the Corn Belt, and atrazine was also the most widely detected pesticide in watersheds in this area, according to the USGS nationwide study. USGS determined that the concentrations of atrazine and other pesticides detected had the potential to adversely affect aquatic plants and invertebrates in some of the streams, since organisms are vulnerable to short-term exposure to relatively small amounts of certain pesticides. Similarly, increased pesticide use for the cultivation of corn could impair groundwater supplies. USGS found pesticides in 61 percent of shallow wells sampled in agricultural areas. Once groundwater is contaminated, it is difficult to clean up, according to the experts we contacted.

According to some of the experts and officials we spoke with, increased demand for biofuel feedstocks may also create incentives for farmers to place marginal lands back into production. Marginal lands generally have lower productivity soils, so cultivating them may require more nutrient and pesticide inputs than more productive lands, potentially leading to further water quality impairments. Furthermore, delivery of sediments, nutrients, and pesticides to surrounding water bodies may increase if these lands are placed back into production because these lands are often highly susceptible to erosion due to wind and water. Of particular concern to many of the experts with whom we spoke are the millions of acres of land currently enrolled in the Conservation Reserve Program (CRP). This federal program provides annual rental payments and cost share assistance to landowners who contractually agree to retire highly erodible or other environmentally-sensitive cropland from agricultural purposes. As part of the contract, farmers are generally required to plant or maintain vegetative covers (such as native grasses) on the land, which provide a range of environmental benefits, including improved water quality, reduced erosion, enhanced wildlife habitat, and preserved soil productivity. However, many experts and officials we spoke with from the five selected states are concerned that higher corn prices and increased demand for biofuel feedstocks may encourage farmers to return CRP land to crop production. If such conversion does occur, these officials noted that water quality may further decline in the future.

Little Is Yet Known about the Water Resource Implications of Next Generation Feedstocks

Next generation feedstocks for biofuels have the potential for fewer negative effects on water resources, although several of the experts and officials that we spoke with said that the magnitude of these effects remains largely unknown because these feedstocks have not yet been grown on a commercial scale. These experts suggested that certain water resource impacts were likely for the following potential feedstocks:

- Agricultural residues, such as corn stover, collected from fields that have already been harvested, can provide feedstock for cellulosic ethanol production. The primary advantage of using agricultural residues is that they are a byproduct of crop cultivation and thus do not require additional water or nutrient inputs. However, removal of these residues has consequences for both soil and water quality, so there may be limits on how much agricultural residues can be removed for cellulosic ethanol production. According to the experts we spoke with, leaving crop residues unharvested on the field benefits soil quality by providing nutrients that help maintain long-term soil productivity, enhancing soil moisture retention, increasing net soil carbon, and reducing the need for nutrient inputs for future crops.[16] In addition, leaving crop residues on the field helps prevent soil erosion due to wind and water and nutrient runoff into the water supply. Farmers could reduce the negative effects of residue removal by harvesting only corn cobs or part of the stover, but the optimal removal rate is not yet fully known, and is currently being studied by several federal agencies and academic institutions.
- Perennial grasses may require less water and provide some water quality benefits. Perennial grasses such as mixed prairie and switchgrass can grow with less water than corn. But some experts cautioned that any water supply benefits from these grasses will only occur if they are rainfed. For instance, officials in Minnesota told us that because the state's crops are primarily rainfed, shifting to the cultivation of

cellulosic feedstocks, like perennial grasses, without irrigation would have a minimal impact on the state's water supply. However, other experts and local officials pointed out that if farmers choose to irrigate perennial grasses in order to achieve maximum yields and profits as they do for other crops, then producing these feedstocks could have the same detrimental effects on water supplies as do other crops. This concern was reiterated by the National Research Council, which stated that while irrigation of native grasses is unusual now, it could easily become more common as cellulosic biofuel production gets under way.[17]

Perennial grasses can also help preserve water quality by reducing soil, nutrient, and pesticide runoff. Research indicates that perennial grasses cycle nitrogen more efficiently than some row crops and protect soil from erosion due to wind and water. As a result, they can reduce the need for most fertilizers after crops are established, and the land on which these crops are grown do not need to be tilled every year, which reduces soil erosion and sedimentation. According to experts, farmers could also plant a mix of perennial grasses, which could minimize the need for pesticides by promoting greater diversity and an abundance of natural enemies for agricultural pests. In addition, perennial grasses cultivated across an agricultural landscape may help reduce nutrient and chemical runoff from farm lands. Grasses can also be planted next to water bodies to help filter out nutrients and secure soil and can serve as a windbreak to help minimize erosion. However, the type of land and cultivation methods used to grow perennial grasses will influence the extent to which they improve water quality. For instance, if perennial grasses were harvested down to the soil, they would not reduce soil erosion as compared to conventional feedstocks in the long run, according to some experts. In addition, according to some experts, if farmers choose to use fertilizers to maximize yields from these crops as they do for other crops or if these crops are grown on lands with decreased soil quality that require increased nutrient application, then cultivation of perennial grasses could also lead to water quality impairments.

- Woody biomass, such as biomass from the thinning of forests and cultivation of certain fast-growing tree varieties, could serve as feedstock for cellulosic ethanol production, according to some experts. Use of thinnings is not expected to impact water supply, as they are residuals from forest management. Thinning of forests can have the added benefit of reducing the intensity of wildfires, the aftermath of which facilitates runoff of nutrients and sediment into surface waters. Waste from urban areas or lumber mills may also provide another source of biomass that would not require additional water resources. This waste would include the woody portions of commercial, industrial, and municipal solid waste, as well as byproducts generated from processing lumber, engineered wood products, or wood particles; however, almost all of the commercial wood waste is currently used as fuels or raw material for existing products. In addition, some experts said that fast-growing tree species, such as poplar, willow, and cottonwood, are potential cellulosic feedstocks. However, these experts also cautioned that some of these varieties may require irrigation to cultivate and may have relatively high consumptive water requirements.
- Algae are also being explored as a possible feedstock for advanced biofuels. According to several experts, one advantage of algae is that they can be cultivated in brackish or degraded water and do not need freshwater supplies. However, currently

algae cultivation is expected to consume a great deal of water, although consumption estimates vary widely—from 40 to 1,600 gallons of water per gallon of biofuel produced, according to experts—depending on what cultivation method is used. With open-air, outdoor pond cultivation, water loss is expected to be greater due to evaporation, and additional freshwater will be needed to replenish the water lost and maintain the water quality necessary for new algal growth. In contrast, when algae are cultivated in a closed environment, as much as 90 percent less water is lost to evaporation, according to one expert.[18]

The Extent to Which Biofuel Conversion May Affect Water Resources also Depends on the Feedstock Used and Biofuel Produced

During the process of converting feedstocks into biofuels, biorefineries not only need a supply of high-quality water, but also discharge certain contaminants that could impact water quality. The amount of water needed and the contaminant discharge vary by type of biofuel produced and type of feedstock used in the conversion process. For example, ethanol production requires greater amounts of high-quality water than does biodiesel. Conversion of corn to ethanol requires approximately 3 gallons of water per gallon of ethanol produced, which represents a decrease from an estimated 5.8 gallons of water per gallon of ethanol in 1998.[19] According to some experts, these gains in efficiency are, for the most part, the result of ethanol plants improving their water recycling efforts and cooling systems.

According to some experts we spoke with, the biofuel conversion process generally requires high-quality water because the primary use for ethanol production is for cooling towers and boilers, and cleaner water transfers heat more efficiently and does less damage to this equipment. As a result, ethanol biorefineries prefer to use groundwater because it is generally cleaner, of more consistent quality, and its supply is less variable than surface water. Furthermore, the use of lesser-quality water leaves deposits on biorefinery equipment that require additional water to remove. However, despite water efficiency gains, some communities have become concerned about the potential impacts of withdrawals for biofuel production on their drinking water and municipal supplies and are pressuring states to limit ethanol facilities' use of the water. For example, at least one Minnesota local water district denied a permit for a proposed biorefinery due to concerns about limited water supply in the area.

Current estimates of the water needed to convert cellulosic feedstocks to ethanol range from 1.9 to 6.0 gallons of water per gallon of ethanol, depending on the technology used. Conversion of these next generation feedstocks is expected to use less water when compared to conventional feedstocks in the long run, according to some experts.[20] For example, officials from a company in the process of establishing a biorefinery expect the conversion of pine and other cellulosic feedstocks to consume less water than the conversion of corn to ethanol once the plant is operating at a commercial scale. However, some researchers cautioned that the processes for converting cellulosic feedstocks currently require greater quantities of water than needed for corn ethanol. They said the technology has not been optimized and commercial-scale production has not yet been demonstrated, therefore any estimates on water use by cellulosic biorefineries are simply projections at this time.

In contrast, biodiesel conversion requires less water than ethanol conversion—approximately 1 gallon of freshwater per gallon of biodiesel.

Similar to ethanol conversion, much of this water is lost during the cooling and feedstock drying processes. Biodiesel facilities can use a variety of plant and animal-based feedstocks, providing more options when choosing a location. This flexibility in type of feedstock that can be converted allows such facilities to be built in locations with plentiful water supplies, lessening their potential impact.

In addition to the water supply effects, biorefineries can have water quality effects because of the contaminants they discharge. However, the type of contaminant discharged varies by the type of biofuel produced. For example, ethanol biorefineries generally discharge chemicals or salts that build up in cooling towers and boilers or are produced as waste by reverse osmosis, a process used to remove salts and other contaminants from water prior to discharge from the biorefinery.[21] EPA officials told us that the concentrated salts discharged from reverse osmosis are a concern due to their effects on water quality and potential toxicity to aquatic organisms. In contrast, biodiesel refineries discharge other pollutants such as glycerin that may be harmful to water quality. EPA officials told us that glycerin from small biodiesel refineries can be a problem if it is released into local municipal wastewater facilities because it may disrupt the microbial processes used in wastewater treatment.[22] Glycerin is less of a concern with larger biodiesel refineries because, according to EPA officials, it is often extracted from the waste stream prior to discharge and refined for use in other products.

Several state officials we spoke with told us these discharges are generally well-regulated under the Clean Water Act. Under the act, refineries that discharge pollutants into federally regulated waters are required to obtain a federal National Pollutant Discharge Elimination System (NPDES) permit, either from EPA or from a state agency authorized by EPA to implement the NPDES program. These permits generally allow a point source, such as a biorefinery, to discharge specified pollutants into federally regulated waters under specific limits and conditions. State officials we spoke with reported they closely monitor the quality of water being discharged from biofuel conversion facilities, and that the facilities are required to treat their water discharges to a high level of quality, sometimes superior to the quality of the water in the receiving water body.

Storage and Distribution of Biofuels Can Have Some Water Quality Consequences

The storage and distribution of ethanol-blended fuels could result in water quality impacts in the event that these fuels leak from storage tanks or the pipes used to transport these fuels. Ethanol is highly corrosive and there is potential for releases into the environment that could contaminate groundwater and surface water, among other issues.[23] When ethanol-blended fuels leak from underground storage tanks (UST) and aboveground tank systems, the contamination may pose greater risks than petroleum. This is because the ethanol in these blended fuels causes benzene, a soluble and carcinogenic chemical in gasoline, to travel longer distances and persist longer in soil and groundwater than it would in the absence of ethanol,[24] increasing the likelihood that it could reach some drinking water supplies.[25] Federal officials told us that, because it is illegal to store ethanol-blended fuels in tanks not designed for the purpose, they had not encountered any concerns specific to ethanol storage. However, officials from two states did express concern about the possibility of leaks and told us that ethanol-blended fuels are still sometimes stored in tanks not designed for the fuel.

For instance, one of these states reported a 700-gallon spill of ethanol-blended fuels due to the scouring of rust plugs in a UST.[26]

According to EPA officials, a large number of the 617,000 federally regulated UST systems currently in use at approximately 233,000 sites across the country are not certified to handle fuel blends that contain more than 10 percent ethanol.[27] Moreover, according to EPA officials, most tank owners do not have records of all the UST systems' components, such as the seals and gaskets. Glues and adhesives used in UST piping systems were not required to be tested for compatibility with ethanol-blended fuel until recently. Thus there may be many compatible tanks used for storing ethanol-blended fuels that have incompatible system components, increasing the potential for equipment failure and fuel leakage, according to EPA officials. EPA told us that it is continuing to work with government and industry partners to study the compatibility of these components with various ethanol blends. EPA officials also stressed the importance of understanding the fate and transport of biofuels into surface water because biofuels are transported mainly by barge, rail, and truck. The officials noted that spills of biofuels or their byproducts have already occurred into surface waters.

The Effect of Increased Biofuel Production Will Vary by Region, Due to Differences in Water Resources and State Laws

According to many experts and officials that we contacted, as biofuel production increases, farmers and the biofuel production industry will need to consider regional differences in water supply and quality when choosing which feedstocks to grow and how and where to expand their biofuel production capacity. Specifically, they noted that in the case of cultivation, certain states may be better suited to cultivate particular feedstocks because of the amount and type of water available. Some examples they provided include the following:

- Certain cellulosic feedstocks, such as switchgrass, would be well-suited for areas with limited rainfall, such as Texas, because these feedstocks generally require less water and are drought tolerant.
- In the Midwest, switchgrass and other native perennial grasses could be grown as stream buffer strips or as cover crops, which are crops planted to keep the soil in place between primary plantings.
- In Georgia, some experts said pine was likely to be cultivated as a next generation biofuel feedstock because the state has relatively limited land available for cultivation and increased cultivation of pine or other woody biomass without irrigation would not cause a strain on water supplies.
- In the Southeast and Pacific Northwest, waste from logging operations and paper production was identified as a potential feedstock for cellulosic ethanol production.
- Areas with limited freshwater supplies and a ready supply of lower-quality water, such as brackish water or water from wastewater treatment plants, would be better suited to the cultivation of algae. For example, Texas was identified as a state

suitable for algae cultivation because of the large amounts of brackish water in many of its aquifers, as well as its abundant sunlight and supplies of carbon dioxide from industrial facilities.

Research indicates that in making decisions about feedstock production for biofuels it will be important to consider the effects that additional cultivation will have on the quality of individual water bodies and regional watersheds. Farmers need to consider local water quality effects when making decisions regarding the suitability of a particular feedstock or where to employ agricultural management practices that minimize nutrient application. In addition, state officials should consider these effects when deciding where programs such as the CRP may be the most effective. For example, experts and officials told us it will be important to identify watersheds in the Midwest that are delivering the largest nutrient loads into the Mississippi River basin and, consequently, contributing to the Gulf of Mexico dead zone, in order to minimize additional degradation that could result from increased crop cultivation in these watersheds. In addition, research has shown it is important that management practices be tailored to local landscape conditions, such as topography and soil quality, and landowner objectives, so that efforts to reduce nutrient and sediment runoff can be maximized.

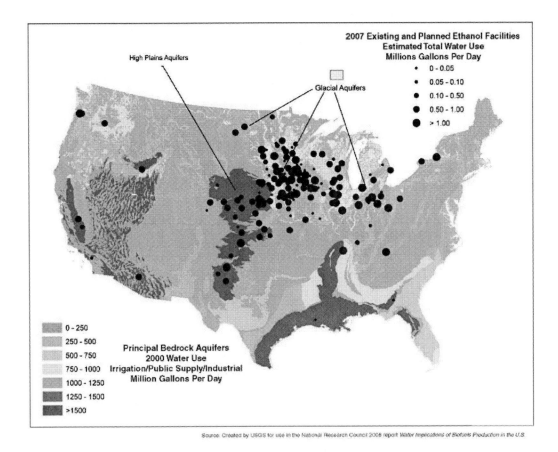

Figure 4. Existing and Planned Ethanol Facilities (as of 2007) and Their Estimated Total Water Use Mapped with the Principal Bedrock Aquifers, including the Ogallala, or High Plains, Aquifer, of the United States and Total Water Use in 2000.

In the case of biofuel conversion, some experts and officials said that state regulators and industry will need to consider the availability of freshwater supplies and the quality of those supplies when identifying and approving sites for biorefineries. Currently, many biorefineries are located in areas with limited water resources. For instance, as figure 4 shows, many existing and planned ethanol facilities are located on stressed aquifers, such as the Ogallala, or High Plains, Aquifer. These facilities require 100,000 to 1 million gallons of water per day, and as mentioned earlier, the rate of water withdrawal from the aquifer is already much greater than its recharge rate, allowing water withdrawals in Nebraska or South Dakota to affect water supplies in other states that draw from that aquifer. Experts noted that states with enough rainfall to replenish underlying aquifers may be more appropriate locations for biorefineries.

Finally, relevant water laws in certain states may influence the location of future biorefineries. Specifically, several states have enacted laws that require permits for groundwater or surface water withdrawals and this requirement could impact where biorefineries will be sited. These laws specify what types of withdrawals must be permitted by the responsible regulatory authority and the requirements for receiving a permit. For instance, Georgia's Environmental Protection Division grants permits for certain withdrawals of groundwater and surface water, including for use by a biorefinery, when the use will not have unreasonable adverse effects on other water uses. According to state officials, there has not yet been a case where a permit for a biorefinery was denied because the amount of projected withdrawal was seen as unreasonable. In contrast, groundwater decisions are made at the local level in Texas, where more than half of the counties have groundwater conservation districts, and Nebraska. In deciding whether to issue a permit, the Texas groundwater conservation districts consider whether the proposed water use unreasonably affects either existing groundwater and surface water resources or existing permit holders, among other factors. In Nebraska, permits are only required for withdrawals and transfers of groundwater for industrial purposes. In addition, in Nebraska, where water supplies are already fully allocated in many parts of the state, natural resource districts can require biofuel conversion facilities to offset the water they will consume by reducing water use in other areas of the region. The volume of withdrawals can also factor into the need for a permit. While Texas conservation district permits are required for almost all types of groundwater wells, Georgia state withdrawal permits are only required for water users who withdraw more than an average of 100,000 gallons per day.[28]

AGRICULTURAL PRACTICES, TECHNOLOGICAL INNOVATIONS, AND ALTERNATIVE WATER SOURCES CAN MITIGATE SOME WATER RESOURCE EFFECTS OF BIOFUELS PRODUCTION, BUT THERE ARE BARRIERS TO ADOPTION

Agricultural conservation practices can reduce the effects of increased biofuel feedstock cultivation on water supply and water quality, but there are several barriers to widespread adoption of these practices. Similarly, the process of converting feedstocks to biofuels, technological innovations, and the use of alternative water sources can help reduce water supply and water quality impacts, but these options can be cost prohibitive and certain noneconomic barriers to their widespread use remain.

Certain Agricultural Practices Can Benefit Water Supply and Water Quality, but Barriers May Limit Widespread Adoption

Many experts and officials we spoke with highlighted the importance of using agricultural conservation practices to reduce the potential effects of increased biofuel feedstock cultivation on water resources. These practices can reduce nutrient and pesticide runoff as well as soil erosion by retaining additional moisture and nutrients in the soil and disturbing the land less. For example, several experts and officials we spoke with said that installing and maintaining permanent vegetation areas adjacent to lakes and streams, known as riparian zones, could significantly reduce the impacts of agricultural runoff. More specifically, several experts and officials said that planting buffer strips of permanent vegetation, such as perennial grasses, or constructing or restoring wetlands in riparian areas would reduce the effects that crop cultivation can have on water quality, as shown in figure 5.

Figure 5. Example of a Riparian Buffer Adjacent to Cropland.

Experts also identified conservation tillage practices—such as "no-till" systems or reduced tillage systems, where the previous year's crop residues are left on the fields and new crops are planted directly into these residues—as an important way to reduce soil erosion (see figure 6). Research conducted by USDA has shown a substantial reduction in cropland erosion since 1985, when incentives were put in place to encourage the adoption of conservation tillage practices.[29] Another practice, crop rotation, also reduces erosion and helps replenish nutrients in the soil. This contrasts with practices such as continuous corn cultivation—in which farmers plant corn on the same land year after year instead of rotating to other crops—which often leads to decreased soil quality. Furthermore, experts identified cover crops, a practice related to crop rotation, as a way to mitigate some of the impacts of agricultural runoff. Cover crops are planted prior to or following a harvested crop, primarily for seasonal soil protection and nutrient recovery before planting the next year's crops. These crops, which include grains or perennial grasses, absorb nutrients and protect the soil surface from erosion caused by wind and rain, especially when combined with conservation tillage practices.

Note: The picture depicts conservation tillage, a process in which last year's crop residues are left on the field and planting occurs directly into this minimally tilled soil.

Figure 6. Example of Conservation Tillage.

Figure 7. Example of Low-Energy Precision-Application Irrigation.

Experts also identified "precision agriculture" as an important tool that can reduce fertilizer runoff and water demand by closely matching nitrogen fertilizer application and irrigation to a crop's nutrient and water needs. Precision agriculture uses technologies such as geographic information systems and global positioning systems to track crop yield, soil moisture content, and soil quality to optimize water and nutrient application rates. Farmers can use this information to tailor water, fertilizer, and pesticide application to specific plots within a field, thus potentially reducing fertilizer and pesticide costs, increasing yields, and reducing environmental impacts. Other precision agriculture tools, like low-energy precision-application irrigation and subsurface drip irrigation systems, operate at lower pressures and have higher irrigation water application and distribution efficiencies than conventional irrigation systems, as shown in figure 7.[30] Several experts and officials said that in order to promote such practices, it is important to continue funding and enrollment in federal programs, such as USDA's Environmental Quality Incentives Program, which pay farmers or provide education and technical support. See appendix II for an expanded discussion of agricultural conservation practices.

Several experts and officials we spoke with also said that genetic engineering has the potential to decrease the water, nutrient, and pesticide requirements of biofuel feedstocks.[31] According to an industry trade group, biotechnology firms are currently developing varieties of drought-resistant corn that may be available to farmers within the next several years. These varieties could significantly increase yields in arid regions of the country that traditionally require irrigation for corn production. Companies are also working to develop crops that absorb additional nutrients or use nutrients more efficiently, giving them the potential to reduce nutrient inputs and the resulting runoff. However, industry officials believe it may be up to a decade before these varieties become available commercially. Furthermore, according to EPA, planting drought-resistant crops, such as corn, may lead to increased cultivation in areas where it has not previously occurred and may result in problems including increased nutrient runoff.

Experts and officials told us there are both economic and noneconomic barriers to the adoption of agricultural conservation practices.

- *Economic barriers.* According to several experts, as with any business, farming decisions are made in an attempt to maximize profits. As a result, experts told us that some farmers may be reluctant to adopt certain conservation practices that may reduce yields and profits, especially in the short term. Furthermore, experts and officials also said that some of these agricultural conservation practices can be costly, especially precision agriculture. For example, the installation of low-energy precision irrigation and subsurface drip irrigation systems is significantly more expensive than conventional irrigation systems because of the equipment needed, among other reasons.[32] Farmers may also hesitate to switch from traditional row crops to next generation cellulosic crops because of potential problems with cash flow and lack of established markets. Specifically, it can take up to 3 years to establish a mature, economically productive crop of perennial grasses, and farmers would be hard-pressed to forgo income during this period. Moreover, farmers may not be willing to cultivate perennial grasses unless they are assured that a market exists for the crop and that they could earn a profit from its cultivation. Furthermore, efficient cultivation and harvest could require farmers to buy new equipment, which would be costly and would add to the price they would have to receive for perennial grasses in order to make a profit.

- *Noneconomic barriers.* Experts and officials we contacted said that many farmers do not have the expertise or training to implement certain practices, and some agricultural practices may be less suited for some places. For example, state officials told us that farmers usually need a year or more of experience with reduced tillage before they can achieve the same crop yields they had with conventional tillage. In addition, precision agriculture relies on technologies and equipment that require training and support. Officials told us that to help address this training need, USDA and states have programs in place that help educate farmers on how to incorporate these practices and, in some cases, provide funding to help do so. In addition, some experts and officials cited regional challenges associated with some agricultural practices and the cultivation of biofuel feedstocks. For example, these experts and officials said that the amount of agricultural residue that can be removed would vary by region and even by farm. Similarly, cultivation of certain cover crops as biofuel feedstocks may not be suitable in the relatively short growing seasons of northern regions.

Use of Innovative Technologies and Alternative Water Sources Could Reduce the Water Resource Effects of Biorefineries, but Costs and Logistics Impede Adoption

Technological improvements have already increased water use efficiency in the ethanol conversion process. Newly built biorefineries with improved processes have reduced water use dramatically over the past 10 years, and some plants have reduced their wastewater discharge to zero. Of the remaining water use, water loss from cooling towers for biorefineries is responsible for approximately 50 to 70 percent of water consumption in modern dry-milling ethanol plants.[33]

Some industry experts we spoke with said that further improvements in water efficiency at corn ethanol plants are likely to come from minimizing water loss from cooling towers or from using alternative water sources, such as effluent from sewage treatment plants. One alternative technology that can substantially reduce water lost through cooling towers is a dry cooling system,[34] which relies primarily on air rather than water to transfer heat from industrial processes.[35] In addition, some ethanol plants are beginning to replace freshwater with alternative sources of water, such as effluent from sewage treatment plants, water from retention ponds at power plants, or excess water from adjacent rock quarries. For example, a corn ethanol conversion plant in Iowa gets a third of its water from a local wastewater treatment plant. By using these alternative water sources, the biorefineries can lower their use of freshwater during the conversion process. While these strategies of improved water efficiency at biorefineries show considerable promise, there are barriers to their adoption. For example, technologies such as dry cooling systems are often prohibitively expensive and can increase energy consumption. Furthermore, according to industry experts, alternative water sources can create a need for expensive wastewater treatment equipment. Some industry experts also told us that the physical layout of a conversion facility may need to be changed to make room for these improvements. Because of the considerable costs of such improvements, several experts told us, it is difficult for biorefineries to integrate these water-conserving technologies while remaining competitive in the economically strained ethanol industry.

Many experts and officials stated that technological innovations for next generation biofuel conversion also have the potential to reduce the water supply and water quality impacts of increased biofuel production. For example, thermochemical production of cellulosic ethanol could require less than 2 gallons of water per gallon of ethanol produced.[36] In addition, some next generation biofuels, known as "drop-in" fuels, are being developed that are compatible with the existing fuel infrastructure, which could reduce the risk that leaks and spills could contaminate local water bodies. For example, biobutanol is produced using fermentation processes similar to those used to make conventional ethanol, but it does not have the same corrosive properties as ethanol and could be distributed through the existing gasoline infrastructure.[37] In addition, liquid hydrocarbons derived from algae have the potential to be converted to gasoline, diesel, and jet fuel, which also can be readily used in the existing fuel infrastructure.[38] However, while these proposed technological innovations can reduce the water resource impacts of increased biofuel production, the efficacy of most of these innovations has not yet been demonstrated on a commercial scale, and some innovations' efficacy has not yet been demonstrated on a pilot scale.

EXPERTS IDENTIFIED A VARIETY OF KEY RESEARCH AND DATA NEEDS RELATED TO INCREASED BIOFUELS PRODUCTION AND LOCAL AND REGIONAL WATER RESOURCES

Many of the experts and officials we spoke with identified areas where additional research is needed to evaluate and understand the effects of increased biofuel production on water resources. These needs fall into two broad areas: (1) research on the water effects of feedstock cultivation and conversion and (2) better data on local and regional water resources.

Experts and officials identified the following research needs on the water resource effects of feedstock cultivation and conversion processes:

Genetically engineered biofuel feedstocks. Many experts and officials cited the need for more research into the development of drought-tolerant and water- and nutrient-efficient crop varieties to decrease the amount of water needed for irrigation and the amount of fertilizer that needs to be applied to biofuel feedstocks. According to the National Research Council, this research should also address the current lack of knowledge on the general water requirements and evapotranspiration rates of genetically engineered crops, including next generation crops.[39] Regarding nutrient efficiency, some experts and officials noted that research into the development of feedstocks that more efficiently take up and store nitrogen from the soil would help reduce nitrogen runoff. In addition, USDA officials added that research to determine the water requirements for conventional biofuel feedstocks and new feedstock varieties developed specifically for biofuel production is also needed.

Effects of cellulosic crops on hydrology. Many experts and officials also told us there is a need to better understand the water requirements of cellulosic crops and the impact of commercial-scale cellulosic feedstock cultivation on hydrology, which is the movement of water through land and the atmosphere into receiving water bodies. According to one expert, these feedstocks differ from corn in their life cycles, root systems, harvest times, and evapotranspiration levels, all of which may influence hydrology. In addition, some research suggests that farmers may cultivate cellulosic feedstocks on marginal or degraded lands because these lands are not currently being farmed and may be suitable for these feedstocks. However, according to the National Research Council, the current evapotranspiration rates of crops grown on such lands is not well known.[40]

Effects of cellulosic crops on water quality. Many experts and officials we spoke with said research is needed to better understand the nutrient needs of cellulosic crops grown on a commercial scale. Specifically, field research is needed on the movement of fertilizer in the soil, air, and water after it is applied to these crops. One expert explained there are water quality models that can describe what happens to fertilizer when applied to corn, soy, and other traditional row crops. However, such models are less precise for perennial grasses due to the lack of data from field trials. Similarly, several experts and officials told us that additional research is also needed on the potential water quality impacts from the harvesting of corn stover. In particular, research is needed on the erosion and sediment delivery rates of different cropping systems in order to determine the acceptable rates of residue removal for different crops, soils, and locations and to develop the technology to harvest residue at these rates.

Cultivation of algae. Although algae can be cultivated using lower-quality water, the impact on water supply and water quality will ultimately depend on which cultivation methods are determined to be the most viable once this nascent technology reaches commercial scale. Many experts we spoke with noted the need for research on how to more efficiently cultivate algae to minimize the freshwater consumption and water quality impacts. For example, research on how to maximize the quantity of water that can be recycled during harvest will be essential to making algae a more viable feedstock option. Further research is also needed to determine whether the pathogens and predators in the lower-quality water are harmful to the algae.[41] In addition, research is also needed on how to manage water discharges during cultivation and harvest of algae. Although it is expected that most water will be recycled, a certain amount must be removed to prevent the buildup of salt. This water may contain pollutants—such as nutrients, heavy metals, and accumulated toxics—that need to be removed to meet federal and state water quality standards.

Data on land use. Better data are needed on what lands are currently being used to cultivate feedstocks, what lands may be most suitable for future cultivation, and how land is actually being managed, according to experts and officials. For example, some experts and officials told us there is a need for improved data on the status and trends in the CRP. According to a CRP official, USDA does not track what happens to land after it is withdrawn from the CRP. Such data would be useful because it would help officials gain a better understanding of the extent to which marginal lands are being put back into production. In addition, improved data on land use would help better target and remove the least productive lands from agricultural production, resulting in water supply and water quality benefits because these lands generally require greater amounts of inputs, according to these experts and officials. Research is also needed to determine optimal placement of feedstocks and use of agricultural conservation practices to get the best yields and minimize adverse environmental impacts.

Farmer decision making. Several experts and officials told us that a better understanding of how farmers make cultivation decisions, such as which crops to plant or how to manage their lands, is needed in the context of the water resource effects of biofuel feedstocks. Specifically, several experts and officials said that research is needed to better understand how farmers decide whether to adopt agricultural conservation practices. In particular, some experts and officials said research should explore how absentee ownership of land affects the choice of farming practices. These experts and officials told us it is common for landowners to live elsewhere and rent their farmland to someone else. For example, in Iowa, 50 percent of agricultural land is rented, according to one expert, and renters may be making cultivation decisions that maximize short-term gains rather than focusing on the long-term health of the land. In addition, several experts and officials said that research is needed to understand the cultural pressures that may make farmers slow to adopt agricultural conservation practices. For example, some experts and officials we spoke with said that some farmers may be hesitant to move away from traditional farming approaches.

Conversion. Existing and emerging technology innovations, such as those discussed earlier in the report, may be able to address some effects of conversion on water resources, but more research into optimizing current technologies is also needed, according to experts. For example, research into new technologies that further reduce water needs for biorefinery cooling systems would have a significant impact on the overall water use at a biorefinery, according to several experts. Congress is considering legislation—the Energy and Water Research Integration Act—that would require DOE's research, development, and demonstration programs to seek to advance energy and energy efficiency technologies that minimize freshwater use, increase water use efficiency, and utilize nontraditional water sources with efforts to improve the quality of that water.[42] It would also require the Secretary of Energy to create a council to promote and enable, in part, improved energy and water resource data collection. Similarly, with regard to conversion facilities for the next generation feedstocks, further research is needed to ensure that the next generation of biorefineries is as water efficient as possible. For example, for the conversion of algae into biofuels, research is needed on how to extract oil from algal cells so as to preserve the water contained in the cell, which would allow some of that water to be recycled.

Storage and distribution. EPA officials noted that additional research related to storage and distribution of biofuels is also needed to help reduce the effects of leaks that can result from the storage of biofuel blends in incompatible tank systems. Although EPA has some research under way, more is needed into the compatibility of fuel blends containing more than 10 percent ethanol with the existing fueling infrastructure. In addition, research should evaluate advanced conversion technologies that can be used to produce a variety of renewable fuels that can be used in the existing infrastructure. Similarly, research is needed into biodiesel distribution and storage, such as assessing the compatibility of blends greater than 5 percent with the existing storage and distribution infrastructure.

In addition, experts and officials identified the following needs for better data on local and regional water resources:

Water availability data. Because some local aquifers and surface water bodies are already stressed, many experts called for more and better data on water resources.[43] Although USGS reports data on water use every 5 years, the agency acknowledges that it does not have good estimates of water use for biofuel production for irrigation or fuel production, so it is unclear how much water has been or will be actually consumed with increases in cultivation and conversion of biofuel feedstocks. Furthermore, some experts and officials told us that even when local water data are available, the data sources are often inconsistent or out of date. For example, the data may capture different information or lack the information necessary for making decisions regarding biofuel production.

According to several experts and officials, better data on water supplies would also help ensure that new biorefineries are built in areas with enough water for current and future conversion processes. Although biorefineries account for only a small percentage of water used during the biofuel production process, the additional withdrawals from aquifers can affect other users that share these water sources. Improving water supply data would help determine whether the existing water supplies can support the addition of a biorefinery in a particular area. Some experts also noted the need for research on the availability of lower-quality water sources such as brackish groundwater, which could be used for cultivation of some next generation feedstocks, especially algae. Better information is necessary to better define the spatial distribution, depth, quantity, physical and chemical characteristics, and sustainable withdrawal rates for these lower-quality water sources, and to predict the long-term effects of water extraction.

Linkages between datasets. Some experts also cited a need for better linkages between existing datasets. For example, datasets on current land use could be combined with aquifer data to help determine what land is available for biofuel feedstock cultivation that would have minimal effects on water resources. In addition, some experts said that while there are data that state agencies and private engineering companies have collected on small local aquifers, a significant effort would be required to identify, coordinate, and analyze this information because linkages do not currently exist.

Geological process data. Several experts and officials also said that research into geological processes is needed to understand the rate at which aquifers are replenished and the impact of increased biofuel production on those aquifers. Although research suggests there should be sufficient water resources to meet future biofuel feedstock production demands at a national level, increased production may lead to significant water shortages in certain regions. For example, additional withdrawals in states relying heavily on irrigation for agriculture may place new demands on already stressed aquifers in the Midwest. Even in water-rich states, such as Iowa, concerns have arisen over the effects of increased biofuel production, and research is needed to assess the hydrology and quality of a state's aquifers to help ensure it is on a path to sustainable production, according to one state official.

AGENCY COMMENTS AND OUR EVALUATION

We provided a draft of this report to USDA, DOE, DOI, and EPA for review and comment. USDA generally agreed with the findings of our report and provided several comments for our consideration. Specifically, USDA suggested that we consider condensing our discussion of agricultural practices, equipment, and grower decisions, as these items may or may not be relevant depending on the feedstock or regulatory control. However, we made no revisions to the report because we believe that cultivation is a significant part of the biofuels life cycle, and these items are relevant and necessary to consider when discussing the potential effect of biofuel production on water resources. USDA also noted that the report is more focused on corn ethanol production than next generation biofuels and that we had not adequately recognized industry efforts to be more sustainable through a movement toward advanced biofuels. Given the maturity of the corn ethanol industry, the extent of knowledge about the effects on water supply and quality from cultivation of corn and its conversion into ethanol, and the uncertainty related to the effects of next generation biofuel production, we believe the balance in the report is appropriate. Moreover, although the shift toward next generation biofuels is a positive step in terms of sustainability, this industry is still developing and the full extent of the environmental benefits from this shift is still unknown. USDA also provided technical comments, which we incorporated as appropriate. See appendix III for USDA's letter.

DOE generally agreed with our findings and approved of the overall content of the report and provided several comments for our consideration. Specifically, DOE noted that it may be too early to make projections on the amount of CRP land that will be converted and the amount of additional inputs that will be needed for cultivation of biofuel feedstocks. In addition, DOE suggested we expand our discussion of efforts to address risks of ethanol transport and note the water use associated with the production of biomass-to-liquid fuels. We adjusted the text as appropriate to reflect these suggestions. DOE also suggested that the report should discuss water pricing; however, this was outside the scope of our review. See appendix IV for DOE's letter.

In its general comments, DOI stated that the report is useful and agreed with the finding on the need for better data on water resources to aid the decision about where to cultivate feedstocks and locate biorefineries. DOI also suggested that the report should include a discussion of the other environmental impacts of biofuel production, such as effects on wildlife habitat or effects on soil. In response, we note that this report was specifically focused on the impacts of biofuel production on water resources; however, for a broader discussion of biofuel production, including other environmental effects, see our August 2009 report.[44] DOI also provided additional technical comments that we incorporated into the report as appropriate. See appendix V for DOI's letter.

EPA did not submit formal comments, but did provide technical comments that we incorporated into the final report as appropriate.

We are sending copies of this report to interested congressional committees; the Secretaries of Agriculture, Energy, and the Interior; the Administrator of the Environmental Protection Agency; and other interested parties. In addition, the report will be available at no charge on the GAO Web site at http://www.gao.gov.

If you or your staff have questions about this report, please contact us at (202) 512-3841 or mittala@gao.gov or gaffiganm@gao.gov. Contact points for our Offices of Congressional Relations and Public Affairs may be found on the last page of this report. GAO staff who made key contributions to this report are listed in appendix VI.

Sincerely yours,

Ms. Anu K. Mittal
Director, Natural Resources and Environment

Mark E. Gaffigan
Director, Natural Resources and Environment

APPENDIX I. OBJECTIVES, SCOPE, AND METHODOLOGY

Our objectives for this review were to describe (1) the known water resource effects of biofuel production in the United States; (2) the agricultural conservation practices and technological innovations that exist or are being developed to address these effects and any barriers that may prevent the adoption of these practices and technologies; and (3) key research needs regarding the effects of biofuel production on water resources.

To address each of these objectives, we conducted a systematic analysis of relevant articles of relevant scientific articles, U.S. multidisciplinary studies, and key federal and state government reports addressing the production of biofuels and its impact on water supply and quality, including impacts from the cultivation of biofuel feedstock and water use and effluent release from biofuel conversion processes. In conducting this review, we searched databases such as SciSearch, Biosis Previews, and ProQuest and used a snowball technique to identify additional studies, asking experts to identify relevant studies and reviewing studies from article bibliographies. We reviewed studies that fit the following criteria for selection: (1) the research was of sufficient breadth and depth to provide observations or conclusions directly related to our objectives; (2) the research was targeted specifically toward projecting or demonstrating effects of increased biofuel feedstock cultivation, conversion, and use on U.S. water supply and water quality; and (3) typically published from 2004 to 2009. We examined key assumptions, methods, and relevant findings of major scientific articles, primarily on water supply and water quality. We believe we have included the key scientific studies and have qualified our findings where appropriate. However, it is important to note that, given our methodology, we may not have identified all of the studies with findings relevant to these three objectives. Where applicable, we assessed the reliability of the data we obtained and found them to be sufficiently reliable for our purposes.

In collaboration with the National Academy of Sciences, we identified and interviewed recognized experts affiliated with U.S.-based institutions, including academic institutions, the federal government, and research-oriented entities. These experts have (1) published research analyzing the water resource requirements of one or more biofuel feedstocks and the implications of increased biofuels production on lands with limited water resources, (2) analyzed the possible effects of increased biofuel production on water, or (3) analyzed the water impacts of biofuels production and use. Together with the National Academy of Sciences' lists of experts, we identified authors of key agricultural and environmental studies as a basis for conducting semistructured interviews to assess what is known about the effects of the increasing production of biofuels and important areas that need additional research. The experts we interviewed included research scientists in such fields as environmental science, agronomy, soil science, hydrogeology, ecology, and engineering.

Furthermore, to gain an understanding of the programs and plans states have or are developing to address increased biofuel production, we conducted in-depth reviews of the following five states: Georgia, Iowa, Minnesota, Nebraska, and Texas. We selected these states based on a number of criteria: ethanol and biodiesel production levels, feedstock cultivation type, reliance on irrigation, geographic diversity among states currently producing biofuels, and approaches to water resource management and law. For each of the states, we analyzed documentation from and conducted interviews with a wide range of stakeholders to gain the views of diverse organizations covering all stages of biofuel production. These stakeholders included relevant state agencies, including those responsible for oversight of agriculture, environmental quality, and water and soil resources; federal agency officials with responsibility for a particular state or region, such as officials from the Department of the Interior's U.S. Geological Survey (USGS), the U.S. Department of Agriculture's (USDA) Natural Resources Conservation Service, and the Environmental Protection Agency (EPA); university researchers; industry representatives; feedstock producers; and relevant nongovernmental organizations, such as state-level corn associations, ethanol producer associations, and environmental organizations. We also conducted site visits to Iowa and Texas to observe agricultural practices and the operation of selected biofuels production plants.

We also interviewed senior officials, scientists, economists, researchers, and other federal officials from USDA, the Departments of Defense and Energy, EPA, the National Aeronautics and Space Administration, the Department of Commerce's National Oceanic and Atmospheric Administration, the National Science Foundation, and USGS about effects on the water supply and water quality during the cultivation of biofuel feedstocks and the conversion and storage of the finished biofuels. In addition, we interviewed state officials from Georgia, Iowa, Minnesota, Nebraska, and Texas as well as agricultural producers and representatives of biofuel conversion facilities to determine the impact of biofuels production in each state. We also interviewed representatives of nongovernmental organizations, such as the Renewable Fuels Association, the Biotechnology Industry Organization, the Pacific Institute, and the Fertilizer Institute.

To conduct the interview content analysis, we reviewed interviews, selected relevant statements from the interviews, and identified and labeled trends using a coding system. Codes were based on trends identified by previous GAO biofuel-related work, background information collected for the review, and the interviews for this review. The methodology for each objective varied slightly, because the first objective focused on regional differences and therefore relied on case study interviews, while analysis performed for the remaining two objectives used expert interviews in addition to case study interviews. Once relevant data were extracted and coded, we used the coded data to identify and analyze trends. For the purposes of reporting our results, we used the following categories to quantify responses of experts and officials: "some" refers to responses from 2 to 3 individuals, "several" refers to responses from 4 to 6 individuals, and "many" refers to responses from 7 or more individuals.

We conducted our work from January 2009 to November 2009 in accordance with all sections of GAO's Quality Assurance Framework that are relevant to our objectives. The framework requires that we plan and perform the engagement to obtain sufficient and appropriate evidence to meet our stated objectives and to discuss any limitations in our work. We believe that the information and data obtained, and the analysis conducted, provide a reasonable basis for any findings and conclusions in this product.

APPENDIX II. EXAMPLES OF AGRICULTURAL PRACTICES AVAILABLE TO REDUCE THE WATER QUALITY AND WATER SUPPLY EFFECTS OF FEEDSTOCK CULTIVATION FOR BIOFUELS

Agricultural conservation practice	Description	Potential environmental benefits
Soil erosion prevention		
Crop residue management	Any tillage method that leaves a portion of the previous crop residues (unharvested portions of the crop) on the soil surface.	• Reduces soil erosion caused by tillage and exposure of bare soil to wind and water. • Reduces water lost to evaporation. • Improves soil quality. • Reduces sediment and fertilizer runoff.
No-till	Method that leaves soil and crop residue undisturbed except for the crop row where the seed is placed in the ground.	• Reduces soil erosion caused by tillage and exposure of bare soil to wind and water. • Reduces water lost to evaporation. • Improves soil quality by improving soil organic matter. • Reduces sediment and fertilizer runoff.
Cover crops	A close-growing crop that temporarily protects the soil during the interim period before the next crop is established.	• Reduces erosion. • Reduces nitrate leaching. • Integrates crops that store nitrogen from the atmosphere (such as soy), replaces the nitrogen that corn and other grains remove from the soil. • Reduces pesticide use by naturally breaking the cycle of weeds, insects, and diseases. • Improves soil quality by improving soil organic matter.
Nutrient pollution reduction		
Crop rotation	Change in the crops grown in a field, usually in a planned sequence. For example, crops could be grown in the following sequence, corn-soy-corn, rather than in continuous corn.	• Integrates crops that obtain nitrogen from the atmosphere (such as soy), replaces the nitrogen that corn and other grains remove from the soil. • Reduces pesticide use by naturally breaking the cycle of weeds, insects, and diseases.
Nutrient management	Use of nutrients to match the rate, timing, form, and application method of fertilizer to crop needs.	• Reduces nutrient runoff and leaching.
Subsurface fertilizer application	Injection of fertilizer below the soil surface.	• Reduces runoff and gaseous emission from nutrients.

(Continued)

Controlled-release fertilizers	Use of fertilizers with water-insoluble coatings that can prevent water-soluble nitrogen from dissolving.	• Reduces nutrient runoff and leaching. • Increases the efficiency of the way nutrients are supplied to and are taken up by the plant, regardless of the crop.
Controlled drainage	Water control structures, such as a flashboard riser, installed in the drainage outlet allow water level to be raised or lowered as needed.	• Minimizes transport of nutrients to surface waters.
Irrigation techniques		
Subsurface drip irrigation systems	Irrigation systems buried directly beneath the crop apply water directly to the root zone.	• Minimizes water lost to evaporation and runoff.
Low-energy precision-application systems	Irrigation systems that operate at lower pressures and have higher irrigation-water application and distribution efficiencies.	• Minimizes net water loss and energy use.
Reclaimed water use	Water recovered from domestic, municipal, and industrial wastewater treatment plants that has been treated to standards that allow safe reuse for irrigation.	• Reduces demand on surface and ground waters.
Multiple benefits		
Wetland restoration	Restoration of a previously drained wetland by filling ditches or removing or breaking tile drains.	• Reduces flooding downstream. • Filters sediment, nutrients, and chemicals. • Provides habitat for wetland plants, amphibians, and birds.
Riparian buffer zones	Strips or small areas of land planted along waterways in permanent vegetation that help control pollutants and promote other environmental benefits.	• Traps sediment. • Filters nutrients. • Provides habitat and corridors for fish and wildlife.
Precision agriculture	A system of management of site-specific inputs (e.g., fertilizer, pesticides) on a site-specific basis such as land preparation for planting, seed, fertilizers and nutrients, and pest control. Precision agriculture may be able to maximize farm production efficiency while minimizing environmental effects. Key technological tools used in this approach include global positioning systems, geog-raphic information systems, real-time soil testing, real-time weather information, etc.	• Reduces nutrient runoff and leaching. • Reduces erosion. • Reduces pesticide use.

Source: GAO analysis.

APPENDIX III. COMMENTS FROM THE U.S. DEPARTMENT OF AGRICULTURE

United States Department of Agriculture
Research, Education, and Economics
Agricultural Research Service

NOV 1 9 2009

2009 NOV 20 PM 3: 37

Ms. Anu Mittal
Government Accountability Office
Director, Natural Resources and Environment
441 G. Street, NW.
Washington, D.C. 20548

Dear Ms. Mittal:

Thank you for the opportunity to review the U.S. Government Accountability Office Draft Report, Energy-Water Nexus: *Many Uncertainties Remain About National and Regional Effects of Increased Biofuel Production on Water Resources* (GAO-10-116).

The Department of Agriculture (USDA) has reviewed the GAO Draft Report and is in general agreement with its findings. We are impressed with its comprehensiveness, the broad scope of topics covered, and accurate assessment of those issues. We agree with the report's general contention regarding the uncertainties of the availability of water resources to sustain increased biofuel production. Moreover, several of the specific issues cited—such as the potential for ground water contamination from benzene production in underground storage tanks (UST)—provide great insight into the many infrastructure deployment challenges currently facing biofuels production and distribution. The report also does an excellent job laying out the parameters that frame the link between bioenergy and water resource management, thus providing an excellent starting point for the establishment of research and development needs to address water availability and quality issues related to increased production of biofuels. Some substantive comments on the report are as follows:

1. We note that some topics stray from the overarching issue of bioenergy production and water management. For instance, the report discusses details of conservation tillage practices, the need for planting and harvesting equipment, and decisions that a grower might or might not make related to production and environmental concerns. Some of these questions may or may not be relevant depending on the biofeedstock to be produced or may be more dependent on regulatory control. Condensing and shortening these areas could improve the focus of the report.

2. The report documents the move toward advanced biofuels development and notes that many producers are beginning to adopt feedstocks that use less water. The report fails to adequately recognize the degree to which industry is already moving along a more sustainable development path.

Ms. Anu Mittal 2

3. The Draft report allocates significant space and attention to grain based ethanol production, even as concentrated efforts and policies are focusing on next generation biofuels.

USDA's technical and specific comments are attached.

Again, thank you for the opportunity to review.

Sincerely,

EDWARD B. KNIPLING
Administrator

Enclosure

APPENDIX IV. COMMENTS FROM THE DEPARTMENT OF ENERGY

Department of Energy
Washington, DC 20585

NOV 1 2 2009   2009 NOV 13 AM 11: 50

Mr. Mark Gaffigan
Director
Natural Resources and Environment
U.S. Government Accountability Office
441 G Street., NW
Washington, DC 20548

Dear Mr. Gaffigan:

Thank you for the opportunity to comment on the draft GAO Report titled: *"Energy-Water Nexus: Many Uncertainties Remain About National and Regional Effects of Increased Biofuel Production on Water Resources"* (GAO-10-116). The Department of Energy (DOE) appreciates the effort put forth by GAO with regard to this report and is in general agreement with GAO's findings and approves of the overall content of the report, but would like to take this occasion to reiterate its assertion that certain sections of the report would benefit from further revision.

First, statements regarding the likely need for additional nutrients and pesticide inputs on marginal lands (page 11) and the role of biofuels in motivating farmers to return Conservation Reserve Program (CRP) land to row crop production (page 12) are speculative. It could be noted that alternative views exist and that it is too early to make projections for CRP conversion and for whether or not additional inputs are needed.

Second, the section on storage and distribution is appropriate but could be expanded. This section would provide a clearer overview of risks of biofuels if they were put into context. The inclusion of a brief description of the risks associated with storing and transporting petroleum products would be a useful comparison to the risks of biofuels storage and distribution. The EIA suggests that the report recognize the dramatic expansion of E10 motor fuel over the past few years and the governmental and industry efforts to address the associated risks of handling ethanol blends. The Department of Transportation and its Pipeline and Hazardous Material Safety Administration (PHMSA) division, in conjunction with industry groups, are engaged in efforts to deal with the associated risks in handling ethanol blends.

Third, it should be noted that in EIA's *Annual Energy Outlook 2009* http://www.eia.doe.gov/oiaf/aeo/index.html projections there is a growing use of biomass-to-liquids (BTL) fuels (5 billion gallons by 2030) to satisfy the EISA 2007 cellulosic biofuels mandate. EIA believes it might be worth mentioning that the production process for BTL requires no continuous water inputs (water is used for cooling but in a closed loop system).

Moreover, pyrolysis oils which are also being currently considered as cellulosic biofuels use no process water as well. The table in the Appendix that summarizes the potential environmental benefits of the agricultural practices is very useful. The inclusion of a similar table summarizing the pros and cons of the various biofuel conversion processes discussed in the text would be a useful addition.

Finally, DOE believes it would be appropriate to raise the issue of price reform for water in this report. Price sends an important signal to consumers. Distorted prices are resulting in overconsumption of water because the full cost of water is not always passed on to the consumer. In areas where water is too inexpensive to monitor, incomplete data on water use exists.

DOE trusts that GAO will consider these suggestions, but does not deem it necessary that the report be revised on account of the three issues raised. Thank you again for the opportunity to comment on the draft Report. We look forward to working with GAO as we continue our efforts to develop the potential of biofuels.

If you have any questions, please contact me or Ms. Martha Oliver, Office of Congressional and Intergovernmental Affairs, at (202) 586-2229.

Sincerely,

Jacques Beaudry-Losique
Deputy Assistant Secretary for Renewable Energy
Office of Technology Development
Energy Efficiency and Renewable Energy

APPENDIX V. COMMENTS FROM THE DEPARTMENT OF THE INTERIOR

United States Department of the Interior
OFFICE OF THE SECRETARY
Washington, DC 20240

NOV 1 2 2009

Ms. Anu Mittal
Director, Natural Resources and Environment
U.S. Government Accountability Office
441 G Street, N.W.
Washington, D.C. 20548

Dear Ms. Mittal:

Thank you for providing the Department of the Interior the opportunity to review and comment on the draft Government Accountability Office report entitled, "*ENERGY-WATER NEXUS: Many Uncertainties Remain about National and Regional Effects of Increased Biofuel Production on Water Resources*" (GAO-10-116).

The GAO report explicitly makes no recommendations; however, we would like to provide technical comments and some general comments. We hope these comments will assist you in preparing the final report. If you have any questions or need additional information, please contact Donna Myers, Chief, National Water-Quality Assessment Program, United States Geological Survey, at (703) 648-5012.

Sincerely,

Anne J. Castle
Assistant Secretary for
Water and Science

Enclosures

End Notes

[1] Pub. L. No. 110-140, § 201 (2007). The act authorizes the Administrator of the Environmental Protection Agency (EPA), in consultation with the Secretaries of Agriculture and Energy, to waive the RFS levels established in the act, by petition or on the Administrator's own motion, if meeting the required level would severely harm the economy or environment of a state, a region, or the United States or there is an inadequate domestic supply. Throughout this report, the RFS levels established in the act are referred to as requirements, even though these levels could be waived by the EPA Administrator.

[2] For additional information on the effects of biofuel production, see GAO, *Biofuels: Potential Effects and Challenges of Required Increases in Production and Use*, GAO-09-446 (Washington, D.C.: Aug. 25, 2009).

[3] Other major sources of freshwater withdrawals in the United States are thermoelectric (39 percent), public water supply (13 percent), and industrial (5 percent) uses. The remaining withdrawals consist of mining (1 percent), domestic (1 percent), aquaculture (1 percent), and livestock (1 percent) uses. S. Hutson et al., "Estimated Use of Water in the United States in 2000," Circular 1268, U.S. Geological Survey (2004).

[4] The RFS applies to transportation fuel sold or introduced into commerce in the 48 contiguous states. However, the Administrator of EPA is authorized, upon a petition from Alaska or Hawaii, to allow the RFS to apply in that state. On June 22, 2007, Hawaii petitioned EPA to opt into the RFS, and the Administrator approved that request. For the purposes of this report, statements that the RFS applies to U.S. transportation fuel refer to the 48 contiguous states and Hawaii.

[5] Pub. L. No. 110-140, § 201 (2007).

[6] Wu, M., M. Mintz, M. Wang, and S. Arora. "Consumptive Water Use in the Production of Ethanol and Petroleum Gasoline," Center for Transportation Research, Energy Systems Division, Argonne National Laboratory (Argonne, Ill., January 2009.)

[7] According to the National Corn Growers Association, across the United States the acres of corn irrigated represent 21 percent of the total irrigated crop area. The volume of water used in corn irrigation represents 7 percent of all irrigation water.

[8] McGuire, V.L., "Water-level changes in the High Plains aquifer, predevelopment to 2007, 2005-2006, and 2006-2007," USGS SIR 2009-5019 (2009).

[9] Maupin, M.A., and Barber, N.L., "Estimated withdrawals from principal aquifers in the United States," USGS Circular 1279 (2000).

[10] Increased corn cultivation could also result in soil erosion, which reduces fertility by reducing nutrient-rich topsoil. It also contributes to sedimentation, which fills channels in deep areas of waterbodies, affecting aquatic life and recreation. Sediment can also carry contaminants, such as fertilizers and pesticides.

[11] The algae themselves do not reduce oxygen; instead, when the algae die, bacteria deplete oxygen as the algae decompose.

[12] Dried distiller's grain, a byproduct of ethanol production used in animal feed, also contains high levels of phosphorous and contributes to overenrichment of surface and marine waters.

[13] Diaz, Robert and Rutger Rosenberg, "Spreading Dead Zones and Consequences for Marine Ecosystems," *Science*, vol. 321 (2008): pp. 926-929.

[14] Alexander, Richard, Richard Smith, Gregory Schwarz, Elizabeth Boyer, Jacqueline Nolan, and John Brakebill, "Difference in Phosphorous and Nitrogen Delivery to the Gulf of Mexico from the Mississippi River Basin," *Environmental Science and Technology*, vol. 42, no. 3 (2008): pp. 822-830.

[15] Gilliom et al., "The Quality of Our Nation's Waters—Pesticides in the Nation's Streams and Ground Water, 1992-2001," USGS Circular 1291 (2006): p. 172.

[16] While some agricultural residues must be left on the ground to maintain soil moisture and carbon content, a significant portion of the total can be removed in many areas. According to a DOE official, in some parts of the country removal of a portion of the residue is needed because the excess residue does not degrade quickly enough and interferes with subsequent crop growth.

[17] National Research Council, *Water Implications of Biofuels Production in the United States*. The National Academies Press, Washington, D.C. (2008).

[18] Water is still lost with closed cultivation due to the cooling needs of the closed systems, among other uses.

[19] In comparison, the recovery and refining of 1 gallon of crude oil requires a total of 3.6 to 7.0 gallons of water. Wu, M. et al., "Consumptive Water Use in the Production of Ethanol and Petroleum Gasoline," Center for Transportation Research, Energy Systems Division, Argonne National Laboratory (Argonne, Ill., January 2009).

[20] DOE's Energy Information Administration's (EIA) *Annual Energy Outlook 2009* projects that there is a sufficient growth in use of biomass-to-liquids (BTL) fuels to meet the EISA cellulosic biofuel requirement and that the production process for BTL fuels does not require continuous water inputs. BTL refers to processes for converting biomass into a range of liquid fuels, such as gasoline and diesel. In addition, EIA noted that certain oils currently eligible for inclusion as cellulosic biofuels also do not use process water.

[21] Reverse osmosis is a filtration process used to purify freshwater by, for example, removing the salts from it. This process is used to treat water prior to discharging it from the ethanol plant.

[22] Glycerin results in elevated levels of biological oxygen demand, which is a measure of how much oxygen it will take to break down the material. According to EPA officials, biodiesel wastewater with small amounts of glycerin and efficient recovery of methanol has a biological oxygen demand of 10,000-15,000 mg/liter, compared to a normal wash water biological oxygen demand of about 200 mg/liter. With glycerin, biodiesel wastewater has a biological oxygen demand of 80,000 mg/liter. Pure glycerin has a biological oxygen demand of 1,000,000 mg/liter.

[23] There are other hazards that may occur from releases of ethanol-blended fuels. For example, some spills of gasoline with ethanol may pose an explosion risk. Large-scale releases of ethanol have been shown to degrade under anaerobic conditions to produce explosive concentrations of methane. According to EPA, this can pose a significant challenge for remediation contractors mitigating biofuel spills. In addition, the methane generated in the subsurface can migrate into overlying buildings, degrading indoor air quality.

[24] When ethanol is present, the ethanol is consumed by micro-organisms in the soil before other, more harmful fuel constituents. This decomposition takes up nutrients and oxygen needed to break down benzene and related compounds. As a result, the benzene plume extends a greater distance.

[25] Mackay, Douglas, Nicholas R. de Sieyes, Murray D. Einarson, Kevin P. Feris, Alexander A. Pappas, Isaac A. Wood, Lisa Jacobson, Larry G. Justice, Mark N. Noske, Kate M. Scow, and John T. Wilson. "Impact of Ethanol on the Natural Attenuation of Benzene, Toluene, and o-Xylene in a Normally Sulfate-Reducing Aquifer." *Environmental Science Technology*, vol. 40 (2006): pp. 6123-6130; and Ruiz-Aguilar, G., K. O'Reilly, and P. Alvarez. "A Comparison of Benzene and Toluene Plume Lengths for Sites Contaminated with Regular vs. Ethanol-Amended Gasoline." *Ground Water Monitoring & Remediation*, vol. 23, no. 1 (winter 2003): pp. 48-53.

[26] EIA noted that use of E10 has dramatically increased over the past few years and that there are governmental and industry efforts, such as the U.S. Department of Transportation's Pipeline and Hazardous Material Safety Administration, that work with industry groups to address risks associated with handling ethanol blends.

[27] Some UST systems are specifically designed to store fuel containing 85 percent ethanol. According to EPA officials, owners using blends containing 85 percent ethanol generally work with a licensed installer to use certified, compatible storage and dispensing equipment. UST systems comprise many components; however, some of these components have not been tested for use with high ethanol fuel blends.

[28] Any entity that withdraws more than 100,000 gallons a day (monthly average) of surface water or 100,000 gallons a day (daily average) of groundwater requires a water permit.

[29] See the USDA-NRCS 2003 Annual National Resources Inventory (http://www.nrcs.usda.gov/technical/NRI/2003/nri03eros-mrb.html).

[30] Low-energy precision-application center-pivot systems discharge water between alternate crop rows planted in a circle. In subsurface drip irrigation, drip tubes are placed from 6 to 12 inches below the soil surface, the depth depending on the soil type, crop, and tillage practices.

[31] In addition to genetically engineering crops, USDA officials commented that traditional breeding techniques offer great potential for decreasing water, nutrient, and pesticide requirements of biofuels feedstocks.

[32] USDA officials noted that use of precision agriculture may also be limited in the cultivation of cellulosic feedstocks due to the costs involved.

[33] Cooling towers are used to control temperatures during the conversion process by transferring the heat to cooler water. This heat is then transferred via evaporation to the atmosphere.

[34] In one type of dry cooling system, steam flows through condenser tubes and is cooled directly by fans blowing air across the outside of these tubes to condense the steam back into liquid water.

[35] GAO, *Energy-Water Nexus: Improvements to Federal Water Use Data Would Increase Understanding of Trends in Power Plant Water Use*, GAO-10-23 (Washington, D.C.: Oct. 16, 2009).

[36] Wu, M. et al., "Consumptive Water Use in the Production of Ethanol and Petroleum Gasoline," Center for Transportation Research, Energy Systems Division, Argonne National Laboratory (Argonne, Ill., January 2009).

[37] Similar to ethanol, biobutanol is an alcohol that can be produced from domestic feedstocks. However, biobutanol has a few advantages over ethanol. Biobutanol has a higher energy content than ethanol and is compatible with the existing infrastructure.

[38] Liquid hydrocarbons, such as petroleum, are a class of chemical compounds containing only hydrogen and carbon. Potentially, hydrocarbons can be derived from substitutes such as oils from plants or algae.

[39] Evapotranspiration refers to the water lost to the atmosphere from soil and water bodies (evaporation) and from plant leaves (transpiration).

[40] National Research Council, *Water Implications of Biofuels Production in the United States*, 2008.

[41] U.S. Department of Energy, "National Algal Biofuels Technology Roadmap," Draft, 2009. In December 2008, DOE convened a workshop to discuss and identify the critical barriers currently preventing the economical production of algal biofuels at a commercial scale. As a result of this workshop, DOE assembled a draft roadmap that highlights a number of areas in need of additional research.

[42] H.R. 3598, 111th Cong. (2009).

[43] The Omnibus Public Land Management Act of 2009 requires, in part, the Secretary of Interior, in coordination with the National Advisory Committee on Water Information and state and local water resource agencies, to establish a national water availability and use assessment program. Pub. L. No. 111-11, § 9508(a) (2009), *codified at* 42 U.S.C. § 10368(a). This program will, among other things, provide a more accurate assessment of the status of the water resources of the United States. The program may address some of the water availability data needs identified by the experts we spoke with.

[44] GAO-09-446.

In: Exploring the Energy-Water Nexus
Editor: Peter D. Wright

ISBN: 978-1-61209-791-6
© 2011 Nova Science Publishers, Inc.

Chapter 4

A BETTER AND COORDINATED UNDERSTANDING OF WATER RESOURCES COULD HELP MITIGATE THE IMPACTS OF POTENTIAL OIL SHALE DEVELOPMENT

United States Government Accountability Office

WHY GAO DID THIS STUDY

Oil shale deposits in Colorado, Utah, and Wyoming are estimated to contain up to 3 trillion barrels of oil—or an amount equal to the world's proven oil reserves. About 72 percent of this oil shale is located beneath federal lands, making the federal government a key player in its potential development. Extracting this oil is expected to require substantial amounts of water and could impact groundwater and surface water. GAO was asked to report on (1) what is known about the potential impacts of oil shale development on surface water and groundwater, (2) what is known about the amount of water that may be needed for commercial oil shale development, (3) the extent to which water will likely be available for commercial oil shale development and its source, and (4) federal research efforts to address impacts to water resources from commercial oil shale development. GAO examined environmental impacts and water needs studies and talked to Department of Energy (DOE), Department of the Interior (Interior), and industry officials.

WHAT GAO RECOMMENDS

GAO recommends that Interior establish comprehensive baseline conditions for water resources in oil shale regions of Colorado and Utah, model regional groundwater movement, and coordinate on water-related research with DOE and state agencies involved in water regulation. Interior generally concurred with GAO's recommendations.

WHAT GAO FOUND

Oil shale development could have significant impacts on the quality and quantity of water resources, but the magnitude of these impacts is unknown because technologies are years from being commercially proven, the size of a future oil shale industry is uncertain, and knowledge of current water conditions and groundwater flow is limited. In the absence of effective mitigation measures, water resources could be impacted from ground disturbances caused by the construction of roads and production facilities; withdrawing water from streams and aquifers for oil shale operations, underground mining and extraction; and discharging waters produced from or used in operations.

Estimates vary widely for the amount of water needed to commercially produce oil shale primarily because of the unproven nature of some technologies and because the various ways of generating power for operations use differing quantities of water. GAO's review of available studies indicated that the expected total water needs for the entire life cycle of oil shale production ranges from about 1 barrel (or 42 gallons) to 12 barrels of water per barrel of oil produced from in-situ (underground heating) operations, with an average of about 5 barrels, and from about 2 to 4 barrels of water per barrel of oil produced from mining operations with surface heating.

Water is likely to be available for the initial development of an oil shale industry, but the size of an industry in Colorado or Utah may eventually be limited by water availability. Water limitations may arise from increases in water demand from municipal and industrial users, the potential of reduced water supplies from a warming climate, fulfilling obligations under interstate water compacts, and the need to provide additional water to protect threatened and endangered fishes.

The federal government sponsors research on the impacts of oil shale on water resources through DOE and Interior. DOE manages 13 projects whose water-related costs total about $4.3 million, and Interior sponsored two water-related projects, totaling about $500,000. Despite this research, nearly all of the officials and experts that GAO contacted said that there are insufficient data to understand baseline conditions of water resources in the oil shale regions of Colorado and Utah and that additional research is needed to understand the movement of groundwater and its interaction with surface water. Federal agency officials also said they seldom coordinate water-related oil shale research among themselves or with state agencies that regulate water. Most officials noted that agencies could benefit from such coordination.

ABBREVIATIONS

BLM	Bureau of Land Management
BOR	Bureau of Reclamation
DOE	Department of Energy
EIS	environmental impact statement
EPA	Environmental Protection Agency
INL	Idaho National Laboratory
Interior	Department of the Interior

NEPA	National Environmental Policy Act
NETL	National Energy Technology Laboratory
OSEC	Oil Shale Exploration Company
OTA	Office of Technology Assessment
PEIS	programmatic environmental impact statement
RD&D	research, development, and demonstration
USGS	U.S. Geological Survey

October 29, 2010

The Honorable Bart Gordon
Chairman
Committee on Science and Technology
House of Representatives

The Honorable Brian N. Baird
Chairman
Subcommittee on Energy and Environment
Committee on Science and Technology
House of Representatives

Being able to tap the vast amounts of oil locked within U.S. oil shale could go a long way toward satisfying the nation's future oil demands. Oil shale is a sedimentary rock containing solid organic material that converts into a type of crude oil when heated. The Green River Formation—an assemblage of over 1,000 feet of sedimentary rocks that lie beneath parts of Colorado, Utah, and Wyoming—contains the world's largest deposits of oil shale. The U.S. Geological Survey (USGS) estimates that the Green River Formation contains about 3 trillion barrels of oil, and about half of this may be recoverable, depending on available technology and economic conditions.[1] This is an amount about equal to the entire world's proven oil reserves. The thickest and richest oil shale within the Green River Formation exists in the Piceance Basin of northwest Colorado and the Uintah Basin of northeast Utah.

The federal government is in a unique position to influence the development of oil shale because 72 percent of the oil shale within the Green River Formation is beneath federal lands managed by the Department of the Interior's (Interior) Bureau of Land Management (BLM). The Department of Energy (DOE) has provided technological and financial support for oil shale development, primarily through its research and development efforts, but oil shale development has been hampered by concerns over potential impacts on the environment, technological challenges, and average oil prices that have been too low to consistently justify investment. In particular, developing oil shale and providing power for oil shale operations and other activities will require large amounts of water—a resource that is already in scarce supply in the arid West where an expanding population is placing additional demands on water. Some analysts project that large scale oil shale development within Colorado could require more water than is currently supplied to over 1 million residents of the Denver metro area and that water diverted for oil shale operations would restrict agricultural and urban development. The potential demand for water is further complicated by the past decade of drought in the West and projections of a warming climate in the future. While there are also other concerns over the impacts from oil shale development, such as impacts to air quality, wildlife, and nearby communities, this report focuses on water impacts.

In response to your request, and building on our two recent reports examining the relationship between other forms of energy production and water use,[2] we examined (1) what is known about the potential impacts of oil shale development on surface water and groundwater, (2) what is known about the amount of water that may be needed for the commercial development of oil shale, (3) the extent to which water will likely be available for commercial oil shale development and its source, and (4) federal research efforts to address impacts on water resources from commercial oil shale development. Our report focuses on oil shale resources within the Green River Formation in the Piceance Basin of northwest Colorado and in the Uintah Basin of northeast Utah because these are the areas in the United States in which the industry is most interested in pursuing oil shale development due to the great thickness and richness of the deposits.

To determine what is known about the potential impacts to surface water and groundwater from commercial oil shale development, we reviewed an environmental impact statement on oil shale development prepared by BLM and various studies from private and public groups. We discussed the completeness and accuracy of these studies in interviews with federal agency officials, state agency personnel involved in regulating water quality and quantity, oil shale industry representatives, and representatives of environmental groups. We also visited oil shale demonstration projects in Colorado. To determine what is known about the amount of water that may be needed for commercial oil shale development, we conducted a comprehensive literature search for studies on water needs, contacted the authors of these studies, and assessed the reasonableness of their estimates. Our review of the literature identified several groups of activities that comprise the life cycle of oil shale production. We then tabulated the water needs identified in each study for each group of activities and expressed the total water needs for the life cycle as a range based on these numbers. To determine the extent to which water is likely to be available for commercial oil shale development and its source, we compared the total needs reflected in this estimated range to the amount of surface water and groundwater that is physically and legally available in the immediate area and to the future demands of municipalities and other industries as projected by federal and state agencies.[3] To review federal research efforts to address the impacts of commercial oil shale development on water resources, we interviewed officials at DOE, the USGS, BLM, and organizations performing the research, including universities and national laboratories, and collected and reviewed relevant documents describing their research. We also discussed areas for future water research as it relates to oil shale with 18 organizations—including the USGS, BLM, the DOE National Energy Technology Laboratory, the DOE Office of Naval Petroleum and Oil Shale Reserves, the U.S. Bureau of Reclamation, three DOE national laboratories, four state regulatory agencies in Colorado and Utah, three water experts, an industry representative, and two universities performing research—to identify gaps in current efforts.

We conducted this performance audit from September 2009 through October 2010 in accordance with generally accepted government auditing standards. These standards require that we plan and perform the audit to obtain sufficient and appropriate evidence to provide a reasonable basis for our findings and conclusions based on our audit objectives. We believe that the evidence obtained provides such a reasonable basis for our findings and conclusions based on our audit objectives.

BACKGROUND

Interest in oil shale as a domestic energy source has waxed and waned since the early 1900s. In 1912, President Taft established an Office of Naval and Petroleum Oil Shale Reserves, and between 1916 and 1924, executive orders set aside federal land in three separate naval oil shale reserves to ensure an emergency domestic supply of oil. The Mineral Leasing Act of 1920 made petroleum and oil shale resources on federal lands available for development under the terms of a mineral lease, but large domestic oil discoveries soon after passage of the act dampened interest in oil shale. Interest resumed at various points during times of generally increasing oil prices. For example, the U.S. Bureau of Mines developed an oil shale demonstration project beginning in 1949 in Colorado, where it attempted to develop a process to extract the oil. The 1970s' energy crises stimulated interest once again, and DOE partnered with a number of energy companies, spawning a host of demonstration projects. Private efforts to develop oil shale stalled after 1982 when crude oil prices fell significantly, and the federal government dropped financial support for ongoing demonstration projects.

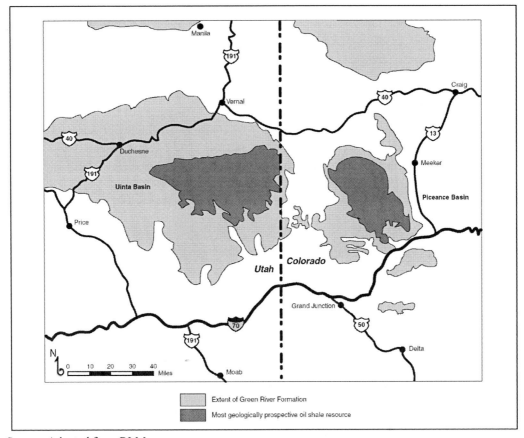

Source: Adopted from BLM.

Figure 1. Location of Oil Shale Resources in Colorado and Utah.

More recently, the Energy Policy Act of 2005 directed BLM to lease its lands for oil shale research and development. In June 2005, BLM initiated a leasing program for research, development, and demonstration (RD&D) of oil shale recovery technologies. By early 2007, it granted six small RD&D leases: five in the Piceance Basin of northwest Colorado and one in Uintah Basin of northeast Utah. The location of oil shale resources in these two basins is shown in figure 1. The leases are for a 10-year period, and if the technologies are proven commercially viable, the lessees can significantly expand the size of the leases for commercial production into adjacent areas known as preference right lease areas. The Energy Policy Act of 2005 directed BLM to develop a programmatic environmental impact statement (PEIS) for a commercial oil shale leasing program. During the drafting of the PEIS, however, BLM realized that, without proven commercial technologies, it could not adequately assess the environmental impacts of oil shale development and dropped from consideration the decision to offer additional specific parcels for lease. Instead, the PEIS analyzed making lands available for potential leasing and allowing industry to express interest in lands to be leased. Environmental groups then filed lawsuits, challenging various aspects of the PEIS and the RD&D program. Since then, BLM has initiated another round of oil shale RD&D leasing and is currently reviewing applications but has not made any awards.

Stakeholders in the future development of oil shale are numerous and include the federal government, state government agencies, the oil shale industry, academic institutions, environmental groups, and private citizens. Among federal agencies, BLM manages the land and the oil shale beneath it and develops regulations for its development. USGS describes the nature and extent of oil shale deposits and collects and disseminates information on the nation's water resources. DOE, through its various offices, national laboratories, and arrangements with universities, advances energy technologies, including oil shale technology. The Environmental Protection Agency (EPA) sets standards for pollutants that could be released by oil shale development and reviews environmental impact statements, such as the PEIS. The Bureau of Reclamation (BOR) manages federally built water projects that store and distribute water in 17 western states and provides this water to users. BOR monitors the amount of water in storage and the amount of water flowing in the major streams and rivers, including the Colorado River, which flows through oil shale country and feeds these projects. BOR provides its monitoring data to federal and state agencies that are parties to three major federal, state, and international agreements, that together with other federal laws, court decisions, and agreements, govern how water within the Colorado River and its tributaries is to be shared with Mexico and among the states in which the river or its tributaries are located. These three major agreements are the Colorado River Compact of 1922, the Upper Colorado River Basin Compact of 1948, and the Mexican Water Treaty of 1944.

The states of Colorado and Utah have regulatory responsibilities over various activities that occur during oil shale development, including activities that impact water. Through authority delegated by EPA under the Clean Water Act, Colorado and Utah regulate discharges into surface waters. Colorado and Utah also have authority over the use of most water resources within their respective state boundaries. They have established extensive legal and administrative systems for the orderly use of water resources, granting water rights to individuals and groups. Water rights in these states are not automatically attached to the land upon which the water is located. Instead, companies or individuals must apply to the state for a water right and specify the amount of water to be used, its intended use, and the specific point from where the water will be diverted for use, such as a specific point on a river or stream. Utah approves the application for a water right through an administrative process, and Colorado approves the application for a water right through a court proceeding. The date of the application establishes its priority—earlier applicants have preferential entitlement to water over later applicants if water availability decreases during a drought. These earlier applicants are said to have senior water rights. When an applicant puts a water right to beneficial use, it is referred to as an absolute water right. Until the water is used, however, the applicant is said to have a conditional water right. Even if the applicant has not yet put the water to use, such as when the applicant is waiting on the construction of a reservoir, the date of the application still establishes priority. Water rights in both Colorado and Utah can be bought and sold, and strong demand for water in these western states facilitates their sale.

Challenges to Oil Shale Development

A significant challenge to the development of oil shale lies in the current technology to economically extract oil from oil shale. To extract the oil, the rock needs to be heated to very high temperatures—ranging from about 650 to 1,000 degrees Fahrenheit—in a process known as retorting. Retorting can be accomplished primarily by two methods. One method involves mining the oil shale, bringing it to the surface, and heating it in a vessel known as a retort. Mining oil shale and retorting it has been demonstrated in the United States and is currently done to a limited extent in Estonia, China, and Brazil. However, a commercial mining operation with surface retorts has never been developed in the United States because the oil it produces competes directly with conventional crude oil, which historically has been less expensive to produce. The other method, known as an in-situ process, involves drilling holes into the oil shale, inserting heaters to heat the rock, and then collecting the oil as it is freed from the rock. Some in-situ technologies have been demonstrated on very small scales, but other technologies have yet to be proven, and none has been shown to be economically or environmentally viable. Nevertheless, according to some energy experts, the key to developing our country's oil shale is the development of an in-situ process because most of the richest oil shale is buried beneath hundreds to thousands of feet of rock, making mining difficult or impossible. Additional economic challenges include transporting the oil produced from oil shale to refineries because pipelines and major highways are not prolific in the remote areas where the oil shale is located and the large-scale infrastructure that would be needed to supply power to heat oil shale is lacking. In addition, average crude oil prices have been lower than the threshold necessary to make oil shale development profitable over time.

Large-scale oil shale development also brings socioeconomic impacts. While there are obvious positive impacts such as the creation of jobs, increase in wealth, and tax and royalty payments to governments, there are also negative impacts to local communities. Oil shale development can bring a sizeable influx of workers, who along with their families, put additional stress on local infrastructure such as roads, housing, municipal water systems, and schools. Development from expansion of extractive industries, such as oil shale or oil and gas, has typically followed a "boom and bust" cycle in the West, making planning for growth difficult. Furthermore, traditional rural uses could be replaced by the industrial development of the landscape, and tourism that relies on natural resources, such as hunting, fishing, and wildlife viewing, could be negatively impacted.

Source: GAO.

Figure 2. Typical View in the Piceance Basin of Colorado.

In addition to the technological, economic, and social challenges to developing oil shale resources, there are a number of significant environmental challenges. For example, construction and mining activities can temporarily degrade air quality in local areas. There can also be long-term regional increases in air pollutants from oil shale processing, upgrading, pipelines, and the generation of additional electricity. Pollutants, such as dust, nitrogen oxides, and sulfur dioxide, can contribute to the formation of regional haze that can affect adjacent wilderness areas, national parks, and national monuments, which can have very strict air quality standards. Because oil shale operations clear large surface areas of topsoil and vegetation, some wildlife habitat will be lost. Important species likely to be negatively impacted from loss of wildlife habitat include mule deer, elk, sage grouse, and raptors. Noise from oil shale operations, access roads, transmission lines, and pipelines can further disturb wildlife and fragment their habitat. In addition, visual resources in the area will be negatively impacted as people generally consider large-scale industrial sites, pipelines, mines, and areas cleared of vegetation to be visually unpleasant (see figure 2 for a typical view within the Piceance Basin). Environmental impacts from oil shale development could be compounded by additional impacts in the area resulting from coal mining, construction, and extensive oil and gas development. Air quality and wildlife habitat appear to be particularly susceptible to the cumulative affect of these impacts, and according to some environmental experts, air quality impacts may be the limiting factor for the development of a large oil shale industry in the future. Lastly, the withdrawal of large quantities of surface water for oil shale operations could negatively impact aquatic life downstream of the oil shale development. Impacts to water resources are discussed in detail in the next section of this report.

OIL SHALE DEVELOPMENT COULD ADVERSELY IMPACT WATER RESOURCES, BUT THE MAGNITUDE OF THESE IMPACTS IS UNKNOWN

Oil shale development could have significant impacts on the quality and quantity of surface and groundwater resources, but the magnitude of these impacts is unknown because some technologies have yet to be commercially proven, the size of a future oil shale industry is uncertain, and knowledge of current water conditions and groundwater flow is limited. Despite not being able to quantify the impacts from oil shale development, hydrologists and engineers have been able to determine the qualitative nature of impacts because other types of mining, construction, and oil and gas development cause disturbances similar to impacts expected from oil shale development. According to these experts, in the absence of effective mitigation measures, impacts from oil shale development to water resources could result from disturbing the ground surface during the construction of roads and production facilities, withdrawing water from streams and aquifers[4] for oil shale operations, underground mining and extraction, and discharging waste waters from oil shale operations.

Quantitative Impacts of Oil Shale Development Cannot Be Measured at This Time

The quantitative impacts of future oil shale development cannot be measured with reasonable certainty at this time primarily because of three unknowns: (1) the unproven nature of in-situ technologies, (2) the uncertain size of a future oil shale industry, and (3) insufficient knowledge of current groundwater conditions. First, geological maps suggest that most of the prospective oil shale in the Uintah and Piceance Basins is more amenable to in-situ production methods rather than mining because the oil shale lies buried beneath hundreds to thousands of feet of rock. Studies have concluded that much of this rock is generally too thick to be removed economically by surface mining, and deep subsurface mines are likely to be costly and may recover no more than 60 percent of the oil shale. Although several companies have been working on the in-situ development of oil shale, none of these processes has yet been shown to be commercially viable. Most importantly, the extent of the impacts of in-situ retorting on aquifers is unknown, and it is uncertain whether methods for reclamation of the zones that are heated will be effective.[5] Second, it is not possible to quantify impacts on water resources with reasonable certainty because it is not yet possible to predict how large an oil shale industry may develop. The size of the industry would have a direct relationship to water impacts. Within the PEIS, BLM has stated that the level and degree of the potential impacts of oil shale development cannot be quantified because this would require making many speculative assumptions regarding the potential of the oil shale, unproven technologies, project size, and production levels. Third, hydrologists at USGS and BLM state that not enough is known about current surface water and groundwater conditions in the Piceance and Uintah Basins. More specifically, comprehensive baseline conditions for surface water and groundwater do not exist. Therefore, without knowledge of current conditions, it is not possible to detect changes in groundwater conditions, much less attribute changes to oil shale development.

A Number of Impacts to Water Quality and Quantity Could Be Expected from Oil Shale Development

Impacts to water resources from oil shale development would result primarily from disturbing the ground surface, withdrawing surface water and groundwater, underground mining, and discharging water from operations.

Ground Disturbances Could Degrade Surface Water Quality

In the absence of effective mitigation measures, ground disturbance activities associated with oil shale development could degrade surface water quality, according to the literature we reviewed and water experts to whom we spoke.[6] Both mining and the in-situ production of oil shale are expected to involve clearing vegetation and grading the surface for access roads, pipelines, production facilities, buildings, and power lines. In addition, the surface that overlies the oil shale would need to be cleared and graded in preparation for mining or drilling boreholes for in-situ extraction. The freshly cleared and graded surfaces would then be exposed to precipitation, and subsequent runoff would drain downhill toward existing gullies and streams. If not properly contained or diverted away from these streams, this runoff could contribute sediment, salts, and possibly chemicals or oil shale products into the nearby streams, degrading their water quality. Surface mining would expose the entire area overlying the oil shale that is to be mined while subsurface mining would expose less surface area and thereby contribute less runoff. One in-situ operation proposed by Shell for its RD&D leases would require clearing of the entire surface overlying the oil shale because wells are planned to be drilled as close as 10 feet apart. Other in-situ operations, like those proposed by American Shale Oil Company and ExxonMobil, envision directionally drilling wells in rows that are far enough apart so that strips of undisturbed ground would remain.[7] The adverse impacts from ground disturbances would remain until exposed surfaces were properly revegetated.

If runoff containing excessive sediment, salts, or chemicals finds its way into streams, aquatic resources could be adversely impacted, according to the water experts to whom we spoke and the literature we reviewed. Although aquatic populations can handle short-term increases in sediment, long-term increases could severely impact plant and animal life. Sediment could suffocate aquatic plants and decrease the photosynthetic activity of these plants. Sediment could also suffocate invertebrates, fish, and incubating fish eggs and adversely affect the feeding efficiency and spawning success of fish. Sedimentation would be exacerbated if oil shale activities destroy riparian vegetation because these plants often trap sediment, preventing it from entering streams. In addition, toxic substances derived from spills, leaks from pipelines, or leaching of waste rock piles could increase mortality among invertebrates and fish.

Surface and underground mining of oil shale will produce waste rock that, according to the literature we reviewed and water experts to whom we spoke, could contaminate surface waters. Mined rock that is retorted on site would produce large quantities of spent shale after the oil is extracted. Such spent shale is generally stored in large piles that would also be exposed to surface runoff that could possibly transport sediment, salts, selenium, metals, and residual hydrocarbons into receiving streams unless properly stabilized and reclaimed. EPA studies have shown that water percolating through such spent shale piles transports pollutants long after abandonment of operations if not properly mitigated. In addition to stabilizing and revegetating these piles, mitigation measures could involve diverting runoff into retention ponds, where it could be treated, and lining the surface below waste rock with impervious materials that could prevent water from percolating downward and transporting pollutants into shallow groundwater. However, if improperly constructed, retention ponds would not prevent the degradation of shallow groundwater, and some experts question whether the impervious materials would hold up over time.

Withdrawing Water for Oil Shale Operations Could Adversely Impact Surface Water and Groundwater

Withdrawing water from streams and rivers for oil shale operations could have temporary adverse impacts on surface water, according to the experts to whom we spoke and the literature we reviewed. Oil shale operations need water for a number of activities, including mining, constructing facilities, drilling wells, generating electricity for operations, and reclamation of disturbed sites. Water for most of these activities is likely to come from nearby streams and rivers because it is more easily accessible and less costly to obtain than groundwater. Withdrawing water from streams and rivers would decrease flows downstream and could temporarily degrade downstream water quality by depositing sediment within the stream channels as flows decrease. The resulting decrease in water would also make the stream or river more susceptible to temperature changes—increases in the summer and decreases in the winter. Elevated temperatures could have adverse impacts on aquatic life, including fishes and invertebrates, which need specific temperatures for proper reproduction and development. Elevated water temperatures would also decrease dissolved oxygen, which is needed by aquatic animals. Decreased flows could also damage or destroy riparian vegetation. Removal of riparian vegetation could exacerbate negative impacts on water temperature and oxygen because such vegetation shades the water, keeping its temperature cooler.

Similarly, withdrawing water from shallow aquifers—an alternative water source—would have temporary adverse impacts on groundwater resources. Withdrawals would lower water levels within these shallow aquifers and the nearby streams and springs to which they are connected. Extensive withdrawals could reduce groundwater discharge to connected streams and springs, which in turn could damage or remove riparian vegetation and aquatic life. Withdrawing water from deeper aquifers could have longer-term impacts on groundwater and connected streams and springs because replenishing these deeper aquifers with precipitation generally takes longer.

Underground Mining and In-Situ Extraction Would Permanently Impact Aquifers

Underground mining would permanently alter the properties of the zones that are mined, thereby affecting groundwater flow through these zones, according to the literature we reviewed and the water experts to whom we spoke. The process of removing oil shale from underground mines would create large tunnels from which water would need to be removed during mining operations. The removal of this water through pumping would decrease water levels in shallow aquifers and decrease flows to streams and springs that are connected. When mining operations cease, the tunnels would most likely be filled with waste rock, which would have a higher degree of porosity and permeability than the original oil shale that was removed.[8] Groundwater flow through this material would increase permanently, and the direction and pattern of flows could change permanently. Flows through the abandoned tunnels could decrease ground water quality by increasing concentrations of salts, metals, and hydrocarbons within the groundwater.

In-situ extraction would also permanently alter aquifers because it would heat the rock to temperatures that transform the solid organic compounds within the rock into liquid hydrocarbons and gas that would fracture the rock upon escape. Water would be cooked off during the heating processes. Some in-situ operations envision using a barrier to isolate thick zones of oil shale with intervening aquifers from any adjacent aquifers and pumping out all the groundwater from this isolated area before retorting.[9] Other processes, like those envisioned by ExxonMobil and AMSO, involve trying to target thinner oil shale zones that do not have intervening aquifers and, therefore, would theoretically not disturb the aquifers. However, these processes involve fracturing the oil shale, and it is unclear whether the fractures could connect the oil shale to adjacent aquifers, possibly contaminating the aquifer with hydrocarbons. After removal of hydrocarbons from retorted zones, the porosity and permeability of the zones are expected to increase, thereby allowing increased groundwater flow. Some companies propose rinsing retorted zones with water to remove residual hydrocarbons. However, the effectiveness of rinsing is unproven, and residual hydrocarbons, metals, salts, and selenium that were mobilized during retorting could contaminate the groundwater. Furthermore, the long-term effects of groundwater flowing through retorted zones are unknown.

Discharge of Waste Waters from Operations Could Temporarily Impact Downstream Waters

The discharge of waste waters from operations would temporarily increase water flows in receiving streams. According to BLM's PEIS, waste waters from oil shale operations that could be discharged include waters used in extraction, cooling, the production of electricity, and sewage treatment, as well as drainage water collected from spent oil shale piles and waters pumped from underground mines or wells used to dewater the retorted zones. Discharges could decrease the quality of downstream water if the discharged water is of lower quality, has a higher temperature, or contains less oxygen. Lower-quality water containing toxic substances could increase fish and invertebrate mortality. Also, increased flow into receiving streams could cause downstream erosion. However, at least one company is planning to recycle waste water and water produced during operations so that discharges and their impacts could be substantially reduced.

ESTIMATES OF WATER NEEDS FOR COMMERCIAL OIL SHALE DEVELOPMENT VARY WIDELY

While commercial oil shale development requires water for numerous activities throughout its life cycle, estimates vary widely for the amount of water needed to commercially produce oil shale. This variation in estimates stems primarily from the uncertainty associated with reclamation technologies for in-situ oil shale development and because of the various ways to generate power for oil shale operations, which use different amounts of water. Based on our review of available information for the life cycle of oil shale production, existing estimates suggest that from about 1 to 12 barrels of water could be needed for each barrel of oil produced from in-situ operations, with an average of about 5 barrels. About 2 to 4 barrels of water could be needed for each barrel of oil produced from mining operations with a surface retort.[10]

Oil Shale Development Requires Water throughout Its Life Cycle

Water is needed for five distinct groups of activities that occur during the life cycle of oil shale development: (1) extraction and retorting, (2) upgrading of shale oil, (3) reclamation, (4) power generation, and (5) population growth associated with oil shale development.

- *Extraction and retorting.* During extraction and retorting, water is used for building roads, constructing facilities, controlling dust, mining and handling ore, drilling wells for in-situ extraction, cooling of equipment and shale oil, producing steam, in-situ fracturing of the retort zones, and preventing fire. Water is also needed for on-site sanitary and potable uses.
- *Upgrading of shale oil.* Water is needed to upgrade, or improve, the quality of the produced shale oil so that it can be easily transported to a refinery. The degree to which the shale oil needs to be upgraded varies according to the retort process. Shale oil produced by surface retorting generally requires more upgrading, and therefore, more water than shale oil produced from in-situ operations that heat the rock at lower temperatures and for a longer time, producing higher-quality oil.
- *Reclamation.* During reclamation of mine sites, water is needed to cool, compact, and stabilize the waste piles of retorted shale and to revegetate disturbed surfaces, including the surfaces of the waste piles. For in-situ operations, in addition to the typical revegetation of disturbed surfaces, as shown in figure 3, water also will be needed for reclamation of the subsurface retorted zones to remove residual hydrocarbons. The volume of water that would be needed to rinse the zones at present is uncertain and could be large, depending primarily on how many times the zones need to be rinsed. In addition, some companies envision reducing water demands for reclamation, as well as for extracting, retorting, and upgrading, by recycling water produced during oil shale operations or by treating and using water produced from nearby oil and gas fields. Recycling technology, however, has not been shown to be commercially viable for oil shale operations, and there could be legal restrictions on using water produced from oil and gas operations.[11]
- *Power generation.* Water is also needed throughout the life cycle of oil shale production for generating electricity from power plants needed in operations. The amount of water used to produce this electricity varies significantly according to generation and cooling technologies employed. For example, thermoelectric power plants use a heat source to make steam, which turns a turbine connected to a generator that makes the electricity. The steam is captured and cooled, often with additional water, and is condensed back into water that is then recirculated through the system to generate more steam. Plants that burn coal to produce steam use more

water for cooling than combined cycle natural gas plants, which combust natural gas to turn a turbine and then capture the waste heat to produce steam that turns a second turbine, thereby producing more electricity per gallon of cooling water. Thermoelectric plants can also use air instead of water to condense the steam. These plants use fans to cool the steam and consume virtually no water, but are less efficient and more costly to run.

- *Population growth.* Additional water would be needed to support an anticipated increase in population due to oil shale workers and their families who migrate into the area. This increase in population can increase the demand for water for domestic uses. In isolated rural areas where oil shale is located, sufficiently skilled workers may not be available.

Estimates of Water Needs for In-Situ Development Vary Significantly

Based on studies that we reviewed, the total amount of water needed for in-situ oil shale operations could vary widely, from about 1 to 12 barrels of water per barrel of oil produced over the entire life cycle of oil shale operations. The average amount of water needed for in-situ oil shale production as estimated by these studies is about 5 barrels. This range is based on information contained primarily in studies published in 2008 and 2009 by ExxonMobil, Shell, the Center for Oil Shale Technology and Research at the Colorado School of Mines, the National Oil Shale Association, and contractors to the state of Colorado.[12] Figure 3 shows Shell's in-situ experimental site in Colorado. Because only two studies examined all five groups of activities that comprise the life cycle of oil shale production, we reviewed water estimates for each group of activities that is described within each of the eight studies we reviewed.[13] We calculated the minimum and the maximum amount of water that could be needed for in-situ oil shale development by summing the minimum estimates and the maximum estimates, respectively, for each group of activities. Differences in estimates are due primarily to the uncertainty in the amount of water needed for reclamation and to the method of generating power for operations.

Source: GAO.

Figure 3. Shell's Experimental In-Situ Site in Colorado.

Table 1. Estimated Barrels of Water Needed for Various Activities per Barrel of Shale Oil Produced by In-Situ Operations

Activity	Minimum estimate	Average estimate	Maximum estimate
Extraction/retorting	0	0.7	1.0
Upgrading liquids	0.6	0.9	1.6
Power generation	0.1	1.5	3.4
Reclamation	0	1.4	5.5
Population growth	0.1	0.3	0.3
Total	**0.8**	**4.8**	**11.8**

Source: GAO analysis of selected studies.

Notes: GAO used from four to six studies to obtain the numbers for each group of activities. See table 8 in appendix I to identify the specific studies. The average for reclamation may be less useful because estimates are either at the bottom or the top of this range.

Table 1 shows the minimum, maximum, and average amounts of water that could be needed for each of the five groups of activities that comprise the life cycle of in-situ oil shale development. The table shows that reclamation activities contribute the largest amount of uncertainty to the range of total water needed for in-situ oil shale operations. Reclamation activities, which have not yet been developed, contribute from 0 to 5.5 barrels of water for each barrel of oil produced, according to the studies we analyzed. This large range is due primarily to the uncertainty in how much rinsing of retorted zones would be necessary to remove residual hydrocarbons and return groundwater to its original quality. On one end of the range, scientists at ExxonMobil reported that retorted zones may be reclaimed by rinsing them several times and using 1 barrel of water or less per barrel of oil produced. However, another study suggests that many rinses and many barrels of water may be necessary. For example, modeling by the Center for Oil Shale Technology and Research suggests that if the retorted zones require 8 or 10 rinses, 5.5 barrels of water could be needed for each barrel of oil produced. Additional uncertainty lies in estimating how much additional porosity in retorted zones will be created and in need of rinsing. Some scientists believe that the removal of oil will double the amount of pore space, effectively doubling the amount of water needed for rinsing. Other scientists question whether the newly created porosity will have enough permeability so that it can be rinsed. Also, the efficiency of recycling waste water that could be used for additional rinses adds to the amount of uncertainty. For example, ExxonMobil scientists believe that almost no new fresh water would be needed for reclamation if it can recycle waste water produced from oil shale operations or treat and use saline water produced from nearby oil and gas wells.

Table 1 also shows that the water needs for generating power contribute significant uncertainty to the estimates of total water needed for in-situ extraction. Estimates of water needed to generate electricity range from near zero for thermoelectric plants that are cooled by air to about 3.4 barrels for coal-fired thermoelectric plants that are cooled by water, according to the studies that we analyzed. These studies suggested that from about 0.7 to about 1.2 barrels of water would be needed if electricity is generated from combined cycle plants fueled by natural gas, depending on the power requirements of the individual oil shale operation. Overall power requirements are large for in-situ operations because of the many electric heaters used to heat the oil shale over long periods of time—up to several years for one technology proposed by industry. However, ExxonMobil, Shell, and AMEC—a contractor to the state of Colorado—believe that an oil shale industry of significant size will not use coal-fired electric power because of its greater water requirements and higher carbon dioxide emissions. In fact, according to an AMEC study, estimates for power requirements of a 1.5 million-barrel-per-day oil shale industry would exceed the current coal-fired generating capacity of the nearest plant by about 12 times, and therefore would not be feasible.[14] Industry representatives with whom we spoke said that it is more likely that a large oil shale industry would rely on natural gas-powered combined cycle thermoelectric plants, with the gas coming from gas fields within the Piceance and Uintah Basins or from gas produced during the retort process. ExxonMobil reports that it envisions cooling such plants with air, thereby using next to no water for generating electricity. However, cooling with air can be more costly and will ultimately require more electricity.

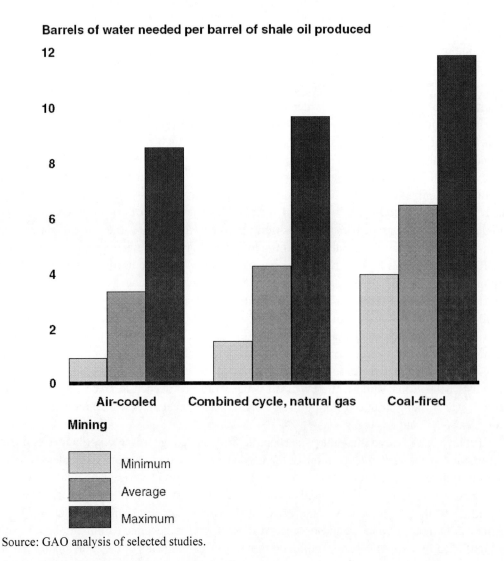

Source: GAO analysis of selected studies.

Figure 4. Estimated Total Barrels of Water Needed per Barrel of Shale Oil Produced by In-Situ Extraction, According to Source of Power Generation.

In addition, table 1 shows that extracting and retorting activities and upgrading activities also contribute to the uncertainty in the estimates of water needed for in-situ operations, but this uncertainty is significantly less than that of reclamation activities or power generation. The range for extraction and retorting is from 0 to 1 barrel of water. The range for upgrading the produced oil is from 0.6 to 1.6 barrels of water, with both the minimum and maximum of this range cited in a National Oil Shale Association study.[15] Hence, each of these two groups of activities contribute about 1 barrel of water to the range of estimates for the total amount of water needed for the life cycle of in-situ oil shale production. Last, table 1 shows there is little variation in the likely estimates of water needed to support the anticipated population increase associated with in-situ oil shale development. Detailed analyses of water needs for population growth associated with an oil shale industry are present in the PEIS, a study by the URS Corporation, and a study completed by the Institute for Clean and Secure Energy at the University of Utah. These estimates often considered the number of workers expected to move into the area, the size of the families to which these workers belong, the ratio of single-family to multifamily housing that would accommodate these families, and per capita water consumption associated with occupants of different housing types.

Figure 4 compares the total water needs over the life cycle of in-situ oil shale development according to the various sources of power generation, as suggested by the studies we reviewed. This is a convenient way to visualize the water needs according to power source. The minimum, average, and maximum values are the sum of the minimum, average, and maximum water needs, respectively, for all five groups of activities. Most of the difference between the minimum and the maximum of each power type is due to water needed for reclamation.

Estimates of Water Needs for Mining and Surface Retorting Vary but Not as Much as the In-Situ Process

Estimates of water needed for mining oil shale and retorting it at the surface vary from about 2 to 4 barrels of water per barrel of oil produced over the entire life cycle of oil shale operations. The average is about 3 barrels of water. This range is based primarily on information obtained through a survey of active oil shale companies completed by the National Oil Shale Association in 2009 and information obtained from three different retorts, as published in a report by the Office of Technology Assessment (OTA) in 1980.[16] Figure 5 shows a surface retort that is operating today at a pilot plant. Because only two studies contained reliable information for all five groups of activities that comprise the life cycle of oil shale production, we reviewed water estimates for each group of activities that is described within each of the eight studies we reviewed.[17] We calculated the minimum and the maximum amount of water that could be needed for mining oil shale by summing the minimum estimates and the maximum estimates, respectively, for each group of activities. The range of water estimates for mining oil shale is far narrower than that of in-situ oil shale production because, according to the studies we reviewed, there are no large differences in water estimates for any of the activities.

Source: Shale Technologies, LLC.

Figure 5. Surface Retort near Rifle, Colorado.

Table 2. Estimated Barrels of Water Needed for Various Activities per Barrel of Shale Oil Produced by Mining and Surface Retorting

Activity	Minimum estimate	Average estimate	Maximum estimate
Extraction/retorting and upgrading liquids	0.9	1.5	1.9
Power generation	0	0.3	0.9
Reclamation	0.6	0.7	0.8
Population growth	0.3	0.3	0.4
Total	**1.8**	**2.8**	**4.0**

Source: GAO analysis of selected studies.
Note: GAO used from three to six studies to obtain the numbers for each group of activities. See table 9 in appendix I to identify the specific studies.

Table 2 shows the minimum, maximum, and average amounts of water that could be needed for each of the groups of activities that comprise the life cycle of oil shale development that relies upon mining and surface retorting. Unlike for in-situ production, we could not segregate extraction and retorting activities from upgrading activities because these activities were grouped together in some of the studies on mining and surface retorting. Nonetheless, as shown in table 2, the combination of these activities contributes 1 barrel of water to the total range of estimated water needed for the mining and surface retorting of oil shale. This 1 barrel of water results primarily from the degree to which the resulting shale oil would need upgrading. An oil shale company representative told us that estimates for upgrading shale oil vary due to the quality of the shale oil produced during the retort process, with higher grades of shale oil needing less processing. Studies in the OTA report did not indicate much variability in water needs for the mining of the oil shale and the handling of ore. Retorts also produce water—about half a barrel for each barrel of oil produced—by freeing water that is locked in organic compounds and minerals within the oil shale. Studies in the OTA report took this produced water into consideration and reported the net anticipated water use.

Table 2 also shows that differences in water estimates for generating power contributed about 1 barrel of water to the range of water needed for mining and surface retorting. We obtained water estimates for power generation either directly from the studies or from power requirements cited within the studies.[18] Estimates of water needed range from zero barrels for electricity coming from thermoelectric plants that are cooled by air to about 0.9 barrels for coal-fired thermoelectric plants that are cooled with water. About 0.3 barrels of water are needed to generate electricity from combined cycle plants fueled by natural gas. Startup oil shale mining operations, which have low overall power requirements, are more likely to use electricity from coal-fired power plants, according to data supplied by oil shale companies, because such generating capacity is available locally. However, a large-scale industry may generate electricity from the abundant natural gas in the area or from gas that is produced during the retorting of oil shale. Water needs for reclamation or for supporting an anticipated increase in population associated with mining oil shale show little variability in the studies that we reviewed.

Figure 6 compares the total water needs over the life cycle of mining and surface retorting of oil shale according to the various sources of power generation. The minimum, average, and maximum values are the sum of the minimum, average, and maximum water needs, respectively, for all five groups of activities.

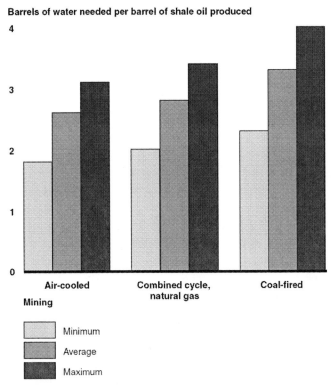

Source: GAO analysis of selected studies.

Figure 6. Estimated Total Barrels of Water Needed per Barrel of Shale Oil Produced by Mining and Surface Retorting, According to Source of Power Generation.

WATER IS LIKELY TO BE AVAILABLE INITIALLY FROM LOCAL SOURCES, BUT THE SIZE OF AN OIL SHALE INDUSTRY MAY EVENTUALLY BE LIMITED BY WATER AVAILABILITY

Water is likely to be available for the initial development of an oil shale industry, but the eventual size of the industry may be limited by the availability of water and demands for water to meet other needs. Oil shale companies operating in Colorado and Utah will need to have water rights to develop oil shale, and representatives from all of the companies with which we spoke are confident that they hold at least enough water rights for their initial projects and will likely be able to purchase more rights in the future. Sources of water for oil shale will likely be surface water in the immediate area, such as the White River, but groundwater could also be used. Nonetheless, the possibility of competing municipal and industrial demands for future water, a warming climate, future needs under existing compacts, and additional water needs for the protection of threatened and endangered fishes, may eventually limit the size of a future oil shale industry.

Oil Shale Companies Own Considerable Water Rights and Options Exist to Obtain More

Companies with interest in oil shale already hold significant water rights in the Piceance Basin of Colorado, and representatives from all of the companies with whom we spoke felt confident that they either had or could obtain sufficient water rights to supply at least their initial operations in the Piceance and Uintah Basins. Western Resource Advocates, a nonprofit environmental law and policy organization, conducted a study of water rights ownership in the Colorado and White River Basins of Colorado and concluded that companies have significant water rights in the area.[19] For example, the study found that Shell owns three conditional water rights[20] for a combined diversion of about 600 cubic feet per second from the White River and one of its tributaries and has conditional rights for the combined storage of about 145,000 acre-feet in two proposed nearby reservoirs.[21] Similarly, the study found that ExxonMobil owns conditional storage capacities of over 161,000 acre-feet on 17 proposed reservoirs in the area. In Utah, the Oil Shale Exploration Company (OSEC), which owns an RD&D lease, has obtained a water right on the White River that appears sufficient for reopening the White River Mine and has cited the possibility of renewing an expired agreement with the state of Utah for obtaining additional water from shallow aquifers connected to the White River. Similarly, Red Leaf Resources cites the possibility of drilling a water well on the state-owned lands that it has leased for oil shale development.

In addition to exercising existing water rights and agreements, there are other options for companies to obtain more water rights in the future, according to state officials in Colorado and Utah. In Colorado, companies can apply for additional water rights in the Piceance Basin on the Yampa and White Rivers. Shell recently applied—but subsequently withdrew the application—for conditional rights to divert up to 375 cubic feet per second from the Yampa River for storage in a proposed reservoir that would hold up to 45,000 acre-feet for future oil shale development. In Utah, however, officials with the State Engineer's office said that additional water rights are not available, but that if companies want additional rights, they could purchase them from other owners. Many people who are knowledgeable on western water rights said that the owners of these rights in Utah and Colorado would most likely be agricultural users, based on a history of senior agricultural rights being sold to developers in Colorado. For example, the Western Resource Advocates study identified that in the area of the White River, ExxonMobil Corporation has acquired full or partial ownership in absolute water rights on 31 irrigation ditches from which the average amount of water diverted per year has exceeded 9,000 acre-feet.[22] These absolute water rights have appropriation dates ranging from 1883 through 1918 and are thus senior to holders of many other water rights, but their use would need to be changed from irrigation or agricultural to industrial in order to be used for oil shale. Also, additional rights may be available in Utah from other sources. According to state water officials in Utah, the settlement of an ongoing legal dispute between the state and the Ute Indian tribe could result in the tribe gaining rights to 105,000 acre-feet per year in the Uintah Basin. These officials said that it is possible that the tribe could lease the water rights to oil shale companies. There are also two water conservancy districts that each hold rights to tens of thousands of acre-feet per year of water in the Uintah Basin that could be developed for any use as determined by the districts, including for oil shale development.

Oil Shale Development Is Likely to Use Local Surface Water, but Groundwater Could Also Be Used

Most of the water needed for oil shale development is likely to come first from surface flows, as groundwater is more costly to extract and generally of poorer quality in the Piceance and Uintah Basins. However, companies may use groundwater in the future should they experience difficulties in obtaining rights to surface water. Furthermore, water is likely to come initially from surface sources immediately adjacent to development, such as the White River and its tributaries that flow through the heart of oil shale country in Colorado and Utah, because the cost of pumping water over long distances and rugged terrain would be high, according to water experts. Shell's attempt to obtain water from the more distant Yampa River shows the importance of first securing nearby sources. In relationship to the White River, the Yampa lies about 20 to 30 miles farther north and at a lower elevation than Shell's RD&D leases. Hence, additional costs would be necessary to transport and pump the Yampa's water to a reservoir for storage and eventual use. Shell withdrew its application citing the global economic downturn.[23]

At least one company has considered obtaining surface water from the even more distant Colorado River, about 30 to 50 miles to the south of the RD&D leases where oil shale companies already hold considerable water rights, but again, the costs of transporting and pumping water would be greater. Although water for initial oil shale development in Utah is also likely to come from the White River as indicated by OSEC's interest, water experts have cited the Green River as a potential water source. However, the longer distance and rugged terrain is likely to be challenging. Figure 7 shows the locations of the oil shale resource areas and their proximity to local surface water sources.

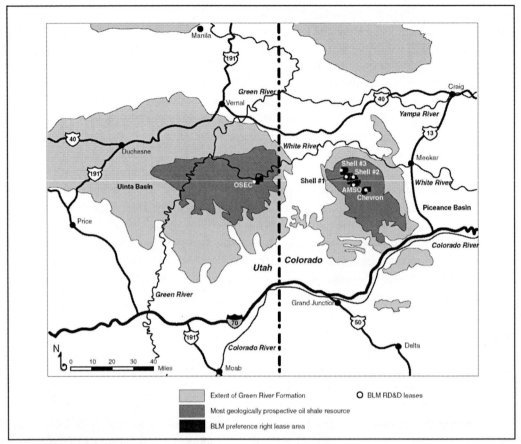

Source: Adopted from BLM.

Figure 7. Location of Rivers near Oil Shale Resources.

In addition to surface water, oil shale companies could use groundwater for operations should more desirable surface water sources be unavailable. However, companies would need to acquire the rights to this groundwater. Shallow groundwater in the Piceance and Uintah Basins occurs primarily within alluvial aquifers, which are aquifers composed of unconsolidated sand and gravel associated with nearby streams and rivers. The states of Utah and Colorado refer to these aquifers legally as tributary waters, or waters that are connected to surface waters and hence are considered to be part of the surface water source when appropriating water rights. Any withdrawal of tributary water is considered to be a withdrawal from the adjacent or nearby stream or river. Less is known about deep groundwater in the Piceance and Uintah Basins, but hydrologists consider it to be of lesser quality, with the water generally becoming increasingly saline with depth. State officials in Utah said that they consider this deeper groundwater to be tributary water, and state officials in Colorado said that they generally consider this deeper water also to be tributary water but will allow water rights applicants to prove otherwise. In the Piceance and Uintah Basins, groundwater is not heavily used, illustrating the reluctance of water users to tap this source. Nevertheless, Shell is considering the use of groundwater, and ExxonMobil is considering using water co-produced with natural gas from nearby but deeper formations in the Piceance Basin. Also, BLM notes that there is considerable groundwater in the regional Bird's Nest

Aquifer in the area surrounding OSEC's RD&D lease in the Uintah Basin. In addition, representatives of oil shale companies said they plan to use water that is released from the organic components of oil shale during the retort process. Since this water is chemically bound within the solid organic components rather than being in a liquid phase, it is not generally viewed as being groundwater, but it is unclear as to how it would be regulated.

Oil Shale Development Will Likely Have to Compete with Increased Demands for Water for Other Needs and a Decreased Water Supply

Developing a sizable oil shale industry may take many years—perhaps 15 or 20 years by some industry and government estimates—and such an industry may have to contend with increased demands for water to meet other needs. Substantial population growth and its correlative demand for water are expected in the oil shale regions of Colorado and Utah. This region in Colorado is a fast-growing area. State officials expect that the population within the region surrounding the Yampa, White, and Green Rivers in Colorado will triple between 2005 and 2050. These officials expect that this added population and corresponding economic growth by 2030 will increase municipal and industrial demands for water, exclusive of oil shale development, by about 22,000 acre-feet per year, or a 76 percent increase from 2000. Similarly in Utah, state officials expect the population of the Uintah Basin to more than double its 1998 size by 2050 and that correlative municipal and industrial water demands will increase by 7,000 acre-feet per year, or an increase of about 30 percent since the mid-1990s. Municipal officials in two communities adjacent to proposed oil shale development in Colorado said that they were confident of meeting their future municipal and industrial demands from their existing senior water rights, and as such will probably not be affected by the water needs of a future oil shale industry. However, large withdrawals could impact agricultural interests and other downstream water users in both states, as oil shale companies may purchase existing irrigation and agricultural rights for their oil shale operations. State water officials in Colorado told us that some holders of senior agricultural rights have already sold their rights to oil shale companies.

Source: GAO.

Figure 8. White River near Meeker, Colorado.

A future oil shale industry may also need to contend with a decreased physical supply of water regionwide due to climate change. A contractor to the state of Colorado ran five projections through a number of climate models and found that their average result suggested that by 2040, a warming climate may reduce the amount of water in the White River in Colorado by about 13 percent, or 42,000 acre-feet. However, there was much variability among the five results, ranging from a 40 percent decrease to a 16 percent increase in today's flow and demonstrating the uncertainty associated with climate predictions. Nevertheless, any decrease would mean that less water would be available downstream in Utah. Because of a warmer climate, the contractor also found that water needed to irrigate crops could increase significantly in the White River Basin, but it is uncertain whether the holders of the water rights used to irrigate the crops would be able to secure this additional water. Simultaneously, the model shows that summer precipitation is expected to decrease, thus putting pressure on farmers to withdraw even more water from local waterways. In addition, the contractor predicted that more precipitation is likely to fall as rain rather than snow in the early winter and late spring. Because snow functions as a natural storage reservoir by releasing water into streams and aquifers as temperatures rise, less snow means that storage and runoff schedules will be altered and less water may be available at different times of the year. Although the model shows that the White River is expected to have reduced flows due to climate change, the same model shows that the Yampa is more likely to experience an increased flow because more precipitation is expected to fall in the mountains, which are its headwaters. Hence, oil shale companies may look to the Yampa for additional water if restrictions on the White are too great, regardless of increased costs to transport the water. While there is not a similar study on climate change impacts for Utah, it is likely that some of the impacts will be similar, considering the close proximity and similar climates in the Uintah and Piceance Basins.

Colorado's and Utah's obligations under interstate compacts could further reduce the amount of water available for development. The Colorado River Compact of 1922, which prescribes how the states through which the Colorado River and its tributaries flow share the river's water, is based on uncharacteristically high flows, as cited in a study contracted by the state of Colorado. Water regulators have since shown that the flow rates used to allocate water under the compact may be 21 percent higher than average historical flow rates, thereby overestimating the amount of water that may be available to share. As a result, the upstream states of Colorado and Utah may not have as much water to use as they had originally planned and may be forced to curtail water consumption so that they can deliver the amount of water that was agreed on in the compact to the downstream states of Arizona, Nevada, and California. Another possible limitation on withdrawals from the Colorado River system is the requirement to protect certain fish species under the Endangered Species Act. Federal officials stated that withdrawals from the Colorado River system, including its tributaries the White and Green Rivers, could be limited by the amount of flow that is necessary to sustain populations of threatened or endangered fishes. Although there are currently no federally mandated minimum flow requirements on the White River in either Utah or Colorado, the river is home to populations of the federally endangered Colorado Pikeminnow, and the Upper Colorado Recovery Program is currently working on a biological opinion which may prescribe minimum flow requirements. In addition, the Green River in Utah is home to populations of four threatened or endangered fishes: the Colorado Pikeminnow, the Razorback Sucker, the Humpback Chub, and the Bonytail Chub. For this reason, agency officials are recommending minimum flow requirements on the Green, which could further restrict the upstream supply of available water.

The Size of an Oil Shale Industry May Be Limited by Water Availability

Although oil shale companies own rights to a large amount of water in the oil shale regions of Colorado and Utah, there are physical and legal limits on how much water they can ultimately withdraw from the region's waterways, and thus limits on the eventual size of the overall industry. Physical limits are set by the amount of water that is present in the river, and the legal limit is the sum of the water that can be legally withdrawn from the river as specified in the water rights held by downstream users. Examining physical limits can demonstrate how much water may be available to all water users. Subtracting the legal limit can demonstrate how much water is available for additional development, providing that current water rights and uses do not change in the future. The state of Colorado refers to this remaining amount of water in the river as that which is physically and legally available.

To put the water needs of a potential oil shale industry in Colorado into perspective, we compared the needs of oil shale industries of various sizes to what currently is physically available in the White River at Meeker, Colorado—a small town immediately east of high-quality oil shale deposits in the Piceance Basin. We also compared the water needs of an oil shale industry to what may be physically and legally available from the White River in 2030. Table 3 shows scenarios depicting the amounts of water that would be needed to develop an oil shale industry of various sizes that relies on mining and surface retorting, based on the studies we examined. Table 4 shows similar scenarios for an oil shale industry that uses in-situ extraction, based on the studies that we examined. The sizes are based on industry and expert opinion and are not meant to be predictions. Both tables assume water demands for peak oil shale production rates, but water use may not follow such a pattern. For example, water use for reclamation activities may not fully overlap with water use for extraction. Also, an industry composed of multiple operations is likely to have some operations at different stages of development. Furthermore, because of the natural variability of stream flows, both on an annual basis and from year-to-year, reservoirs would need to be built to provide storage, which could be used to release a consistent amount of water on a daily basis.

Data maintained by the state of Colorado indicate the amount of water that is physically available in the Whiter River at Meeker, Colorado, averages about 472,000 acre-feet per year.[24] Table 3 suggests that this is much more water than is needed to support the water needs for all the sizes of an industry relying on mining and surface retorting that we considered. Table 4, however, shows that an industry that uses in-situ extraction could be limited just by the amount of water physically available in the White River at Meeker, Colorado. For example, based on an oil shale industry that uses about 12 barrels of water for each barrel of shale oil it produces, such an industry could not reach 1 million barrels per day if it relied solely on physically available water from the White River.

Table 3. Estimated Water Needs for Mining and Surface Retorting of Oil Shale by Industries of Various Sizes

Size of industry (barrels of oil per day)	Minimum waterneeds (acre-feetper year)[a]	Average waterneeds (acre-feetper year)[b]	Maximum waterneeds (acre-feetper year)[c]
25,000[d]	2,400	3,500	4,700
50,000[e]	4,700	7,100	9,400
75,000	7,100	10,600	14,100
100,000	9,400	14,100	18,800
150,000[f]	14,100	21,200	28,200

Source: GAO analysis of selected studies on water needs.

[a] This scenario assumes 2 barrels of water are needed to produce 1 barrel of shale oil. All figures are rounded to the nearest 100 acre-feet.
[b] This scenario assumes 3 barrels of water are needed to produce 1 barrel of shale oil.
[c] This scenario assumes 4 barrels of water are needed to produce 1 barrel of shale oil.
[d] URS, the contractor to the state of Colorado, used this level as the minimum size for a mining operation with a surface retort.
[e] Several literature sources and oil shale companies cite this level as a reasonable commercial operation.
[f] GAO estimated industry size based on three operations of 50,000 barrels per day each.

Table 4. Estimated Water Needs for In-Situ Retorting of Oil Shale by Industries of Various Sizes

Size of industry(barrels of oil perday)	Minimum water needs (acre-feet per year)[a]	Average water needs (acre-feet per year)[b]	Maximum water needs (acre-feet per year)[c]
500,000[d]	24,000	118,000	282,000
1,000,000	47,000	235,000	565,000
1,500,000	71,000	353,000	847,000
2,000,000	94,000	470,000	1,129,000
2,500,000[e]	118,000	588,000	1,411,000

Source: GAO analysis of selected studies on water needs.

[a] This scenario assumes 1 barrel of water is needed to produce 1 barrel of shale oil. All figures are rounded to the nearest 100 acre-feet.
[b] This scenario assumes 5 barrels of water are needed to produce 1 barrel of shale oil.
[c] This scenario assumes 12 barrels of water are needed to produce 1 barrel of shale oil.
[d] One oil shale company with whom we spoke estimated that an oil shale industry could grow to this level, based on analogy to oil sands being developed in Alberta, Canada.
[e] DOE uses this level as the high end for its size estimates of an oil shale industry.

Comparing an oil shale industry's needs to what is physically and legally available considers the needs of current users and the anticipated needs of future users, rather than assuming all water in the river is available to an oil shale industry. The amount of water that is physically and legally available in the White River at Meeker is depicted in table 5. According to the state of Colorado's computer models, holders of water rights downstream use on average about 153,000 acre-feet per year, resulting in an average of about 319,000 acre-feet per year that is currently physically and legally available for development near Meeker. By 2030, however, the amount of water that is physically and legally available is expected to change because of increased demand and decreased supply. After taking into account an anticipated future decrease of 22,000 acre-feet per year of water due to a growing population, about 297,000 acre-feet per year may be available for future development if current water rights and uses do not change by 2030. However, there may be additional decreases in the amount of physically and legally available water in the White River due to climate change, demands under interstate agreements, and water requirements for threatened or endangered fishes, but we did not include these changes in table 5 because of the large uncertainty associated with estimates.

Comparing the scenarios in table 4 to the amount of water that is physically and legally available in table 5 shows the sizes that an in-situ oil shale industry may reach relying solely on obtaining new rights on the White River. The scenarios in table 4 suggest that if an in-situ oil shale industry develops to where it produces 500,000 barrels of oil per day—an amount that some experts believe is reasonable—an industry of this size could possibly develop in Colorado even if it uses about 12 barrels of water per barrel of shale oil it produces. Similarly, the scenarios suggest that an in-situ industry that uses about 5 barrels of water per barrel of oil produced—almost the average from the studies in which power comes from combined cycle natural gas plants—could grow to 1 million barrels of oil per day using only the water that appears to be physically and legally available in 2030 in the White River. Table 4 also shows that an industry that uses just 1 barrel of water per barrel of shale oil produced could grow to over 2.5 million barrels of oil per day.

Regardless of these comparisons, more water or less water could be available in the future because it is unlikely that water rights will remain unchanged until 2030. For example, officials with the state of Colorado reported that conditional water rights—those rights held but not used—are not accounted for in the 297,000 acre-feet per year of water that is physically and legally available because holders of these rights are not currently withdrawing water. These officials also said that the amount of conditional water rights greatly exceeds the flow in the White River near Meeker, and if any of these conditional rights are converted to absolute rights and additional water is then withdrawn downstream, even less water will be available for future development. However, officials with the state of Colorado said that some of these conditional water rights are already owned by oil shale companies, making it unnecessary for some companies to apply for new water rights. In addition, they said, some of the absolute water rights that are accounted for in the estimated 153,000 acre-feet per year of water currently being withdrawn are already owned by oil shale companies. These are agricultural rights that were purchased by oil shale interests who leased them back to the original owners to continue using them for agricultural purposes. Should water not be available from the White River, companies would need to look to groundwater or surface water outside of the immediate area.

There are less data available to predict future water supplies in Utah's oil shale resource area. The state of Utah did not provide us summary information on existing water rights held by oil shale companies. According to the state of Colorado, the average annual physical flow of the White River near the Colorado-Utah border is about 510,000 acre-feet per year. Any amount withdrawn from the White River in Colorado would be that much less water that would be available for development downstream in Utah. The state of Utah estimates that the total water supply of the Uintah Basin, less downstream obligations under interstate compacts, is 688,000 acre-feet per year.[25] Much of the surface water contained in this amount is currently being withdrawn, and water rights have already been filed for much of the remaining available surface water.

Table 5. Estimated Water That Will Be Physically and Legally Available in the White River at Meeker, Colorado, in 2030

	Acre-feet per year
Average historic flow, or water that is physically available today	472,000
Average water use by holders of downstream water rights	-153,000
Average physically and legally available water today	319,000
Estimated increase in municipal and industrial use by 2030	-22,000
Estimated physically and legally available supply in 2030	297,000

Source: GAO analysis of state of Colorado data.

FEDERAL RESEARCH EFFORTS ON THE IMPACTS OF OIL SHALE DEVELOPMENT ON WATER RESOURCES DO NOT PROVIDE SUFFICIENT DATA FOR FUTURE MONITORING

Although the federal government sponsors research on the nexus between oil shale development and water, a lack of comprehensive data on the condition of surface water and groundwater and their interaction limit efforts to monitor the future impacts of oil shale development. Currently DOE funds some research related to oil shale and water resources, including research on water rights, water needs, and the impacts of oil shale development on water quality. Interior also performs limited research on characterizing surface and groundwater resources in oil shale areas and is planning some limited monitoring of water resources. However, there is general agreement among those we contacted—including state personnel who regulate water resources, federal agency officials responsible for studying water, water researchers, and water experts—that this ongoing research is insufficient to monitor and then subsequently mitigate the potential impacts of oil shale development on water resources. In addition, DOE and Interior officials noted that they seldom formally share the information on their water-related research with each other.

DOE Funds Research on Water Rights, Water Needs, and the Impacts of Oil Shale Development on Water Resources

DOE has sponsored most of the oil shale research that involves water-related issues. This research consists of projects managed by the National Energy Technology Laboratory (NETL), the Office of Naval Petroleum and Oil Shale Reserves, and the Idaho National Laboratory. As shown in table 6, DOE has sponsored 13 of 15 projects initiated by the federal government since June 2006. DOE's projects account for almost 90 percent of the estimated $5 million[26] that is to be spent by the federal government on water-related oil shale research through 2013.[27] Appendix II contains a list and description of these projects.

NETL sponsors the majority of the water-related oil shale research currently funded by DOE. Through workshops, NETL gathers information to prioritize research. For example, in October 2007, NETL sponsored the Oil Shale Environmental Issues and Needs Workshop that was attended by a cross-section of stakeholders, including officials from BLM and state water regulatory agencies, as well as representatives from the oil shale industry. One of the top priorities that emerged from the workshop was to develop an integrated regional baseline for surface water and groundwater quality and quantity. As we have previously reported, after the identification of research priorities, NETL solicits proposals and engages in a project selection process.[28] We identified seven projects involving oil shale and water that NETL awarded since June 2006. The University of Utah, Colorado School of Mines, the Utah Geological Survey, and the Idaho National Laboratory (INL) are performing the work on these projects. These projects cover topics such as water rights, water needs for oil shale development, impacts of retorting on water quality, and some limited groundwater modeling. One project conducted by the Colorado School of Mines involves developing a geographic information system for storing, managing, analyzing, visualizing, and disseminating oil shale data from the Piceance Basin. Although this project will provide some baseline data on surface water and groundwater and involves some theoretical groundwater modeling, the project's researchers told us that these data will neither be comprehensive nor complete. In addition, NETL-sponsored research conducted at the University of Utah involves examining the effects of oil shale processing on water quality, new approaches to treat water produced from oil shale operations, and water that can be recycled and reused in operations.

INL is sponsoring and performing research on four water-related oil shale projects while conducting research for NETL and the Office of Naval Petroleum and Oil Shale Reserves. The four projects that INL is sponsoring were self-initiated and funded internally through DOE's Laboratory Directed Research and Development program. Under this program, the national laboratories have the discretion to self-initiate independent research and development, but it must focus on the advanced study of scientific or technical problems, experiments directed toward proving a scientific principle, or the early analysis of experimental facilities or devices. Generally, the researchers propose projects that are judged by peer panels and managers for their scientific merits. An INL official told us they selected oil shale and water projects because unconventional fossil fuels, which include oil shale, are a priority in which they have significant expertise.

Table 6. Federal Funding for Oil Shale Research Initiated Since June 2006

Sponsoring office	Number of oil shale research projects	Federal share of funding for all oil shale research projects	Number of water-related projects	Federal share of funding for water-related projects
DOE National Energy Technology Lab	13	$15,424,702	7	$2,433,097
DOE Office of Naval Petroleum and Oil Shale Reserves	2	2,468,000	2	920,000
DOE Idaho National Lab	5	3,012,500[a]	4	965,000[a]
BLM	3	535,000	2	520,000
USGS	1	1,100,000	0	0
Total	24	$22,540,202	15	$4,838,097

Source: GAO analysis of DOE and Interior data.
[a] Numbers may contain some nonfederal funds.

According to DOE officials, one of the projects managed by the Office of Naval Petroleum and Oil Shale Reserves is directed at research on the environmental impacts of unconventional fuels. The Los Alamos National Laboratory is conducting the work for DOE, which involves examining water and carbon-related issues arising from the development of oil shale and other unconventional fossil fuels in the western United States. Key water aspects of the study include the use of an integrated modeling process on a regional basis to assess the amounts and availability of water needed to produce unconventional fuels, water storage and withdrawal requirements, possible impacts of climate change on water availability, and water treatment and recycling options. Although a key aspect of the study is to assess water availability, researchers on the project told us that little effort will be directed at assessing groundwater, and the information developed will not result in a comprehensive understanding of the baseline conditions for water quality and quantity.

Interior Funds Limited Oil Shale-Related Research on Groundwater and Surface Water Resources and Monitoring

Within Interior, BLM is sponsoring two oil shale projects related to water resources with federal funding totaling about $500,000.[29] The USGS is conducting the research for both projects. For one of the projects, which is funded jointly by BLM and a number of Colorado cities and counties plus various oil shale companies, the research involves the development of a common repository for water data collected from the Piceance Basin. More specifically, the USGS has developed a Web-based repository of water quality and quantity data obtained by identifying 80 public and private databases and by analyzing and standardizing data from about half of them. According to USGS officials, many data elements are missing, and the current repository is not comprehensive. The second project, which is entirely funded by BLM, will monitor groundwater quality and quantity within the Piceance Basin in 5 existing wells and 10 more to be determined at a future date. Although USGS scientists said that this is a good start to understanding groundwater resources, it will not be enough to provide a regional understanding of groundwater resources.

Gaps in Groundwater and Surface Water Data Have Been Identified by Federal and State Officials

Federal law and regulations require the monitoring of major federal actions, such as oil shale development. Regulations developed under the National Environmental Policy Act (NEPA)[30] for preparing an environmental impact statement (EIS), such as the EIS that will be needed to determine the impacts of future oil shale development, require the preparing agency to adopt a monitoring and enforcement program if measures are necessary to mitigate anticipated environmental impacts.[31] Furthermore, the *NEPA Task Force Report to the Council on Environmental Quality* noted that monitoring must occur for long enough to determine if the predicted mitigation effects are achieved.[32] The council noted that monitoring and consideration of potential adaptive measures to allow for midcourse corrections, without requiring new or supplemental NEPA review, will assist in accounting for unanticipated changes in environmental conditions, inaccurate predictions, or subsequent information that might affect the original environmental conditions. In September 2007, the Task Force on Strategic Unconventional Fuels—an 11-member group that included the Secretaries of DOE and Interior and the Governors of Colorado and Utah—issued a report with recommendations on promoting the development of fuels from domestic unconventional fuel resources as mandated by the Energy Policy Act of 2005. This report included recommendations and strategies for developing baseline conditions for water resources and monitoring the impacts from oil shale development. It recommended that a monitoring plan be developed and implemented to fill data gaps at large scales and over long periods of time and to also develop, model, test, and evaluate short- and long-term monitoring strategies. The report noted that systems to monitor water quality would be evaluated; additional needs would be identified; and relevant research, development, and demonstration needs would be recommended.

Also in September 2007, the USGS prepared for BLM a report to improve the efficiency and effectiveness of BLM's monitoring efforts.[33] The report noted that regional water-resources monitoring should identify gaps in data, define baseline conditions, develop regional conceptual models, identify impacts, assess the linkage of impacts to energy development, and understand how impacts propagate. The report also noted that in the Piceance Basin, there is no local, state-level, or national comprehensive database for surface water and groundwater data. Furthermore, for purposes of developing a robust and cost-effective monitoring plan, the report stated that a compilation and analysis of available data are necessary. One of the report's authors told us that the two BLM oil shale projects that the USGS is performing are the initial steps in implementing such a regional framework for water resource monitoring. However, the author said that much more work is needed because so much water data are missing. He noted the current data repository is not comprehensive and much more data would be needed to determine whether oil shale development will create adverse effects on water resources.

Nearly all the federal agency officials, state water regulators, oil shale researchers, and water experts with whom we spoke said that more data are needed to understand the baseline condition of groundwater and surface water, so that the potential impacts of oil shale development can be monitored (see appendix I for a list of the agencies we contacted). Several officials and experts to whom we spoke stressed the need to model the movement of groundwater and its interaction with surface water to understand the possible transport of contaminants from oil shale development. They suggested that additional research would help to overcome these shortcomings. Specifically, they identified the following issues:

- *Insufficient data for establishing comprehensive baseline conditions for surface water and groundwater quality and quantity.* Of the 18 officials and experts we contacted, 17 noted that there are insufficient data to understand the current baseline conditions of water resources in the Piceance and Uintah Basins. Such baseline conditions include the existing quantity and quality of both groundwater and surface water. Hydrologists among those we interviewed explained that more data are needed on the chemistry of surface water and groundwater, properties of aquifers, age of groundwater, flow rates and patterns of groundwater, and groundwater levels in wells. Although some current research projects have and are collecting some water data, officials from the USGS, Los Alamos National Laboratory, and the universities doing this research agreed their data are not comprehensive enough to support future monitoring efforts. Furthermore, Colorado state officials told us that even though much water data were generated over time, including during the last oil shale boom, little of these data have been assimilated, gaps exist, and data need to be updated in order to support future monitoring.

- *Insufficient research on groundwater movement and its interaction with surface water for modeling possible transport of contaminants.* Sixteen of 18 officials and experts to whom we spoke noted that additional research is needed to develop a better understanding of the interactions between groundwater and surface water and of groundwater movement. Officials from NETL explained that this is necessary in order to monitor the rate and pattern of flow of possible contaminants resulting from the in-situ retorting of oil shale. They noted that none of the groundwater research currently under way is comprehensive enough to build the necessary models to understand the interaction and movement. NETL officials noted more subsurface imaging and visualization are needed to build geologic and hydrologic models and to study how quickly groundwater migrates. These tools will aid in monitoring and providing data that does not currently exist.

Interior and DOE Officials Generally Have Not Shared Information on Oil Shale Research

Interior and DOE officials generally have not shared current research on water and oil shale issues. USGS officials who conduct water-related research at Interior and DOE officials at NETL, which sponsors the majority of the water and oil shale research at DOE, stated they have not talked with each other about such research in almost 3 years. USGS staff noted that although DOE is currently sponsoring most of the water-related research, USGS researchers were unaware of most of these projects. In addition, staff at Los Alamos National Laboratory who are conducting some water-related research for DOE noted that various researchers are not always aware of studies conducted by others and stated that there needs to be a better mechanism for sharing this research. Based on our review, we found there does not appear to be any formal mechanism for sharing water-related research activities and results among Interior, DOE, and state regulatory agencies in Colorado and Utah. The last general meeting to discuss oil shale research among these agencies was in October 2007, although there have been opportunities to informally share research at the annual Oil Shale Symposium, the last one of which was conducted at the Colorado School of Mines in October 2010. Of the various officials with the federal and state agencies, representatives from research organizations, and water experts we contacted, 15 of 18 noted that federal and state agencies could benefit from collaboration with each other on water-related research involving oil shale. Representatives from NETL, who are sponsoring much of the current research, stated that collaboration should occur at least every 6 months.

We and others have reported that collaboration among government agencies can produce more public value than one agency acting alone.[34] Specifically concerning water resources, we previously reported that coordination is needed to enable monitoring programs to make better use of available resources in light of organizations often being unaware of data collected by other groups.[35] Similarly in 2004, the National Research Council concluded that coordination of water research is needed to make deliberative judgments about the allocation of funds, to minimize duplication, to present to Congress and the public a coherent strategy for federal investment, and to facilitate large-scale multiagency research efforts.[36] In 2007, the Subcommittee on Water Availability and Quality within the Office of Science and Technology Policy, an office that advises the President and leads interagency efforts related to science and technology stated, "Given the importance of sound water management to the Nation's well-being it is appropriate for the Federal government to play a significant role in providing information to all on the status of water resources and to provide the needed research and technology that can be used by all to make informed water management decisions."[37] In addition, H.R. 1145—the National Water Research and Development Initiative Act of 2009—which has passed the House of Representatives and is currently in a Senate committee, would establish a federal interagency committee to coordinate all federal water research, which totals about $700 million annually. This bill focuses on improving coordination among agency research agendas, increasing the transparency of water research budgeting, and reporting on progress toward research outcomes.

CONCLUSIONS

The unproven nature of oil shale technologies and choices in how to generate the power necessary to develop this resource cast a shadow of uncertainty over how much water is needed to sustain a commercially viable oil shale industry. Additional uncertainty about the size of such an industry clouds the degree to which surface and groundwater resources could be impacted in the future. Furthermore, these uncertainties are compounded by a lack of knowledge of the current baseline conditions of groundwater and surface water, including their chemistry and interaction, properties of aquifers, and the age and rate of movement of groundwater, in the arid Piceance and Uintah Basins of Colorado and Utah, where water is considered one of the most precious resources. All of these uncertainties pose difficulties for oil shale developers, federal land managers, state water regulators, and current water users in their efforts to protect water resources.

Attempts to commercially develop oil shale in the United States have spanned nearly a century. During this time, the industry has focused primarily on overcoming technological challenges and trying to develop a commercially viable operation. More recently, the federal government has begun to focus on studying the potential impacts of oil shale development on surface water and groundwater resources. However, these efforts are in their infancy when compared to the length of time that the industry has spent on attempting to overcome technological challenges. These nascent efforts do not adequately define current baseline conditions for water resources in the Piceance and Uintah Basins, nor have they begun to model the important interaction of groundwater and surface water in the region. Thus they currently fall short of preparing federal and state governments for monitoring the impacts of any future oil shale development. In addition, there is a lack of coordination among federal agencies on water-related research and a lack of communicating results among themselves and to the state regulatory agencies. Without such coordination and communication, federal and state agencies cannot begin to develop an understanding of the potential impacts of oil shale development on water resources and monitor progress toward shared water goals. By taking steps now, the federal government, working in concert with the states of Colorado and Utah, can position itself to help monitor western water resources should a viable oil shale industry develop in the future.

RECOMMENDATIONS FOR EXECUTIVE ACTION

To prepare for possible impacts from the future development of oil shale, we are making three recommendations to the Secretary of the Interior. Specifically, the Secretary should direct the appropriate managers in the Bureau of Land Management and the U.S. Geological Survey to

1. establish comprehensive baseline conditions for groundwater and surface water quality, including their chemistry, and quantity in the Piceance and Uintah Basins to aid in the future monitoring of impacts from oil shale development in the Green River Formation;
2. model regional groundwater movement and the interaction between groundwater and surface water, in light of aquifer properties and the age of groundwater, so as to help in understanding the transport of possible contaminants derived from the development of oil shale; and
3. coordinate with the Department of Energy and state agencies with regulatory authority over water resources in implementing these recommendations, and to provide a mechanism for water-related research collaboration and sharing of results.

AGENCY COMMENTS

We provided a copy of our draft report to Interior and DOE for their review and comment. Interior provided written comments and generally concurred with our findings and recommendations. Interior highlighted several actions it has under way to begin to implement our recommendations. Specifically, Interior stated that with regard to our first recommendation to establish comprehensive baseline conditions for surface water and groundwater in the Piceance and Uintah Basins, implementation of this recommendation includes ongoing USGS efforts to analyze existing water quality data in the Piceance Basin and ongoing USGS efforts to monitor surface water quality and quantity in both basins. Interior stated that it plans to conduct more comprehensive assessments in the future. With regard to our second recommendation to model regional groundwater movement and the interaction between groundwater and surface water, Interior said BLM and USGS are working on identifying shared needs for modeling. Interior underscored the importance of modeling prior to the approval of large-scale oil shale development and cites the importance of the industry's testing of various technologies on federal RD&D leases to determine if production can occur in commercial quantities and to develop an accurate determination of potential water uses for each technology. In support of our third recommendation to coordinate with DOE and state agencies with regulatory authority over water resources,

Interior stated that BLM and USGS are working to improve such coordination and noted current efforts with state and local authorities. Interior's comments are reproduced in appendix III.

DOE also provided written comments, but did not specifically address our recommendations. Nonetheless, DOE indicated that it recognizes the need for a more comprehensive and integrated cross-industry/government approach for addressing impacts from oil shale development. However, DOE raised four areas where it suggested additional information be added to the report or took issue with our findings. First, DOE suggested that we include in our report appropriate aspects of a strategic plan drafted by an ad hoc group of industry, national laboratory, university, and government representatives organized by the DOE Office of Naval Petroleum and Oil Shale Reserves. We believe aspects of this strategic plan are already incorporated into our report. For example, the strategic plan of this ad hoc group calls for implementing recommendations of the Task Force on Strategic Unconventional Fuels, which was convened by the Secretary of Energy in response to a directive within the Energy Policy Act of 2005. The Task Force on Strategic and Unconventional fuels recommended developing baseline conditions for water resources and monitoring the impacts from oil shale development, which is consistent with our first recommendation. The ad hoc group's report recognized the need to share information and collaborate with state and other federal agencies, which is consistent with our third recommendation. As such, we made no changes to this report in response to this comment.

Second, DOE stated that we overestimated the amount of water needed for in-situ oil shale development and production. We disagree with DOE's statement because the estimates presented in our report respond to our objective, which was to describe what is known about the amount of water that may be needed for commercial oil shale development, and they are based on existing publicly available data. We reported the entire range of reputable studies without bias to illustrate the wide range of uncertainty in water needed to commercially develop oil shale, given the current experimental nature of the process. We reported only publicly available estimates based on original research that were substantiated with a reasonable degree of documentation so that we could verify that the estimates covered the entire life cycle of oil shale development and that these estimates did not pertain solely to field demonstration projects, but were instead scalable to commercial operations. We reviewed and considered estimates from all of the companies that DOE identified in its letter. The range of water needed for commercial in-situ development of oil shale that we report ranges from 1 to 12 barrels of water per barrel of oil. These lower and upper bounds represent the sum of the most optimistic and most pessimistic estimates of water needed for all five groups of activities that we identified as comprising the life cycle of in-situ oil shale development. However, the lower estimate is based largely on estimates by ExxonMobil and incorporates the use of produced water, water treatment, and recycling, contrary to DOE's statement that we dismissed the significance of these activities. The upper range is influenced heavily by the assumption that electricity used in retorting will come from coal-fired plants and that a maximum amount of water will be used for rinsing the retorted zones, based on modeling done at the Center for Oil Shale Technology and Research.[38] The studies supporting these estimates were presented at the 29th Annual Oil Shale Symposium at the Colorado School of Mines. Such a range overcomes the illusion of precision that is conveyed by a single point estimate, such as the manner in which DOE cites the 1.59 barrels of water from the AMEC study, or the bias associated with reporting a narrow range based on the assumption that certain technologies will prevail before they are proven to be commercially viable for oil shale development. Consequently, we made no changes to the report in response to this comment.

Third, DOE stated that using the amount of water in the White River at Meeker, Colorado, to illustrate the availability of water for commercial oil shale development understates water availability. We disagree with DOE's characterization of our illustration. The illustration we use in the report is not meant to imply that an entire three-state industry would be limited by water availability at Meeker. Rather, the illustration explores the limitations of an in-situ oil shale industry only in the Piceance Basin. More than enough water appears available for a reasonably sized industry that depends on mining and surface retorting in the Piceance basin. Our illustration also suggests that there may be more than enough water to supply a 2.5 million barrel-per-day in-situ industry at minimum water needs, even considering the needs of current water users and the anticipated needs of future water users. In addition, the illustration suggests that there may be enough water to supply an in-situ industry in the Piceance Basin of between 1 and 2 million barrels per day at average water needs, depending upon whether all the water in the White River at Meeker is used or only water that is expected to be physically and legally available in the future. However, the illustration does point out limitations. It suggests that at maximum water needs, an in-situ industry in the Piceance Basin may not reach 1 million barrels per day if it relied solely on water in the White River at Meeker. Other sources of water may be needed, and our report notes that these other sources could include water in the Yampa or Colorado Rivers, as well as groundwater. Use of produced water and recycling could also reduce water needs as noted in the draft report. Consequently, we made no changes to the report in response to this comment.

Fourth, DOE stated that the report gives the impression that all oil shale technologies are speculative and proving them to be commercially viable will be difficult, requiring a long period of time with uncertain outcomes. We disagree with this characterization of our report. Our report clearly states that there is uncertainty regarding the commercial viability of in-situ technologies. Based on our discussions with companies and review of available studies, Shell is the only active oil shale company to have successfully produced shale oil from a true in-situ process. Considering the uncertainty associated with impacts on groundwater resources and reclamation of the retorted zone, commercialization of an in-situ process is likely to be a number of years away. To this end, Shell has leased federal lands from BLM to test its technologies, and more will be known once this testing is completed. With regard to mining oil shale and retorting it at the surface, we agree that it is a relatively mature process. Nonetheless, competition from conventional crude oil has inhibited commercial oil shale development in the United States for almost 100 years. Should some of the companies that DOE mentions in its letter prove to be able to produce oil shale profitably and in an environmentally sensitive manner, they will be among the first to overcome such long-standing challenges. We are neither dismissing these companies, as DOE suggests, nor touting their progress. In addition, it was beyond the scope of our report to portray the timing of commercial oil shale production or describe a more exhaustive history of oil shale research, as DOE had recommended, because much research currently is privately funded and proprietary. Therefore, we made no changes to the report in response to this comment. DOE's comments are reproduced in appendix IV.

As agreed with your office, unless you publicly announce the contents of this report earlier, we plan no further distribution until 30 days from the report date. At that time, we will send copies of this report to the appropriate congressional committees, Secretaries of the Interior and Energy, Directors of the Bureau of Land Management and U.S. Geological Survey, and other interested parties. In addition, the report will be available at no charge on GAO's Web site at http://www.gao.gov.

If you or your staff have any questions about this report, please contact one of us at (202) 512-3841 or gaffiganm@gao.gov or mittala@gao.gov. Contact points for our Offices of Congressional Relations and Public Affairs may be found on the last page of this report. Key contributors to this report are listed in appendix V.

Mark Gaffigan
Director, Natural Resources and Environment

Anu Mittal
Director, Natural Resources and Environment

APPENDIX I. SCOPE AND METHODOLOGY

To determine what is known about the potential impacts to groundwater and surface water from commercial oil shale development, we reviewed the *Proposed Oil Shale and Tar Sands Resource Management Plan Amendments to Address Land Use Allocations in Colorado, Utah, and Wyoming and Final Programmatic Environmental Impact Statement (PEIS)* prepared by the Bureau of Land Management in September 2008. We also reviewed environmental assessments prepared on Shell Oil's plans for in-situ development of its research, demonstration, and development (RD&D) tracts in Colorado and on the Oil Shale Exploration Company's (OSEC) plan to mine oil shale on its RD&D tract in Utah because these two companies have made the most progress toward developing in-situ and mining technologies, respectively. In addition, we reviewed the Office of Technology Assessment's (OTA) 1980 report, *An Assessment of Oil Shale Technologies;* the Rand Corporation's 2005 report, *Oil Shale Development in the United States;* and the Argonne National Laboratory's 2005 report, *Potential Ground Water and Surface Water Impacts from Oil Shale and Tar Sands Energy-Production Operations.* Because the PEIS was the most comprehensive of these documents, we summarized impacts to groundwater and surface water quantity and quality described within this document and noted that these impacts were entirely qualitative in nature and that the magnitude of impacts was indeterminate because the in-situ technologies have yet to be developed. To confirm these observations and the completeness of impacts within the PEIS, we contacted the Environmental Protection Agency, the Colorado Division of Water Resources, the Colorado Water Conservation Board, the Division of Water Quality within the Colorado Department of Public Health and Environment, the Utah Division of Water Resources, the Utah Division of Water Quality, and the Utah Division of Water Rights—all of which have regulatory authority over some aspect of water resources. To ensure that we identified the range of views on the potential impacts of oil shale development on groundwater and surface water, we also contacted the U.S. Geological Survey (USGS), the Colorado Geological Survey, the Utah Geological Survey, industry representatives, water experts, and numerous environmental groups for their views on the impacts of oil shale on water resources. To assess the impacts of oil shale development on aquatic resources, we reviewed the PEIS and contacted the Colorado Division of Wildlife and the Utah Division of Wildlife Resources.

To determine what is known about the amount of water that may be needed for commercial oil shale development, we searched the Internet and relevant databases of periodicals using the words "oil shale" together with "water use." We also searched Web sites maintained by the Bureau of Land Management (BLM), USGS, and the Department of Energy (DOE) for information on oil shale and water use and interviewed officials at these agencies to determine if there were additional studies that we had not identified. We also checked references cited within the studies for other studies. We limited the studies to those published in 1980 or after because experts with whom we consulted either considered the studies published before then to be adequately summarized in OTA's 1980 report or to be too old to be relevant. We included certain data within the OTA report because some of the surface retort technologies are similar to technologies being tested today. We did not consider verbal estimates of water needs unless companies could provide more detailed information. The 17 studies that we identified appear in table 7.

For further analysis, we divided the studies into two major groups—in-situ extraction and mining with a surface retort. We dismissed a combination of mining and in-situ extraction because most of these technologies are more than 30 years old and generally considered to be infeasible today. The single company that is pursuing such a combination of technologies today—Red Leaf Resources— has not published detailed data on water needs. After reviewing these studies, we found that most of the studies did not examine water needs for the entire life cycle of oil shale development. As such, we identified logical groups of activities based on descriptions within the studies. We identified the following five groups of activities: (1) extraction and retorting, (2) generating power, (3) upgrading shale oil, (4) reclamation, and (5) population growth associated with oil shale development. We did not include refining because we believe it is unlikely that oil shale production will reach levels in the near- or midterm to justify building a new refinery.

To characterize the water needs for the entire life cycle of oil shale development, we identified within each study the water needs for each of the five groups of activities. Except for OTA's 1980 report, which is now 30 years old, we contacted the authors of each study and discussed the estimates with them. If estimates within these studies were given for more than one group of activities, we asked them to break down this estimate into the individual groups when possible. We only considered further analyzing water needs for groups of activities that were based on original research so as not to count these estimates multiple times.

Table 7. Studies on Water Use for Oil Shale Development Initially Identified by GAO

Bartis, et al. *Oil Shale Development in the United States: Prospects and Policy Issues*. Rand Corporation, 2005.
Boak, Jeremy and Earl Mattson. *Water Use for In-Situ Production of Shale Oil from the Green River Formation*. Presented at the 29th Oil Shale Symposium, Colorado School of Mines, October 20, 2009.
Bureau of Land Management. *Oil Shale Research, Development, and Demonstration Project*. Environmental Assessment CO-110-2006-117-EA. Prepared to analyze a proposal by Shell Frontier Oil and Gas, Inc., 2006.
Bureau of Land Management. *Oil Shale Research, Development and Demonstration Project*, White River Mine, Uintah County, Utah. Environmental Assessment UT-080-06-280. Prepared to analyze a proposal by the Oil Shale Exploration Company, April 2007.
Bureau of Land Management. *Proposed Oil Shale and Tar Sands Resource Management Plan Amend-ments to Address Land Use Allocations in Colorado, Utah, and Wyoming and Final Programmatic Environmental Impact Statement (PEIS)*. September 2008.
Dudley-Murphy, Beth, et al. *Meeting Data Needs to Perform a Water Impact Assessment for Oil Shale Development in the Uinta and Piceance Basins, Appendix D in Utah Heavy Oil Program: Final Scientific/Technical Report*. Institute for Clean and Secure Energy, October 2009.
Harding, Benjamin. AMEC Earth and Environmental. *Energy Development Water Needs Assessment and Water Supply Alternatives Analysis*. Presented at the Promise and Perils of Oil Shale Symposium sponsored by the Natural Resources Law Center at the University of Colorado at Boulder, February 5, 2010.
Mangmeechai, Aweewan. *Life Cycle Greenhouse Gas Emissions, Consumptive Water Use and Levelized Costs of Unconventional Oil in North America*. Ph.D. dissertation, Carnegie Mellon University, August 2009.
Mangmeechai, Aweewan et al. *Life Cycle Consumptive Water Use of U.S. Oil Shale*. Presented at the International Society for Industrial Ecology, Boston, Massachusetts, September 29-October 2, 2009.
National Oil Shale Association. "NOSA Evaluates Oil Shale Water Usage." *Oil Shale Update*, vol. II, issue I (September 2009).
Office of Technology Assessment. *An Assessment of Oil Shale Technologies*. June 1980.
Shell Frontier Oil and Gas, Inc. Plan of Operations, *Oil Shale Test Project*. February 15, 2006.
Thomas, Michele Mosio, et al. ExxonMobil Upstream Research. *Responsible Development of Oil Shale*. Presented at the 29th Oil Shale Symposium, Colorado School of Mines, October 2009.
URS Corporation. *Energy Development Water Needs Assessment (Phase I Report)*. Glenwood Springs, Colorado, September 2008.
Veil, J. A. and M.G. Puder. *Potential Ground Water and Surface Water Impacts from Oil Shale and Tar Sands Energy-Production Operations*, Argonne National Lab. October 2006.
Western Resource Advocates. *Water on the Rocks: Oil Shale Water Rights in Colorado*. Boulder, Colorado, 2009.
Wilson, C, et al. Los Alamos National Laboratory. *Assessment of Climate Variability on Water Resource Availability for Oil Shale Development*. Presented at the First Western Forum on Energy and Water Sustainability, School of Environmental Science and Management, University of California, Santa Barbara, March 22-23, 2007.

Source: GAO.

Note: While this table includes all the studies we initially identified, we describe further in this section of the report how we identified data within these studies that sufficiently met our quality criteria to be included in the range of water estimates.

For example, original research on water needs for extraction and retorting may have analyzed mine plans, estimated water needs for drilling wells, estimated water needs for dust control, and discussed recycling of produced water. Original research on water needs for population growth may have discussed the number of workers immigrating to a region, their family size, per capita water consumption, and the nature of housing required by workers. On the other hand, estimates of water needs that were not based on original research generally reported water needs for multiple groups of activities in barrels of water per barrel of oil produced and cited someone else's work as the source for this number. We excluded several estimates that seemed unlikely. For example, we eliminated a water estimate for power generation that included building a nuclear power plant and water estimates for population growth where it was assumed that people would decrease their water consumption by over 50 percent. We also excluded technologies developed prior to 1980 that are dissimilar to technologies being considered by oil shale companies today. We checked mathematical calculations and reviewed power requirements and the reasonableness of associated water needs. For power estimates that did not include associated water needs, we converted power needs into water needs using 480 gallons per megawatt hour of electricity produced by coal-fired, wet recirculating thermoelectric plants and 180 gallons per megawatt hour of electricity produced by gas-powered, combined cycle, wet recirculating thermoelectric plants. Air-cooled systems consume almost no water for cooling. Where appropriate, we also estimated shale oil recoveries based the company's estimated oil shale resources and estimated water needs for rinsing retorted zones based on anticipated changes to the reservoir.

Table 8. Studies GAO Examined That Contained Original Research on Water Requirements for Groups of Activities Representing the Complete Life Cycle for the In-Situ Production of Oil Shale[a]

Study	Extraction And retorting	Power	Reclamation	Upgrading liquids	Population growth
BLM, PEIS					X
Dudley-Murphy et al., table 6, scenarios 2 and 6[b]					X
NOSA, report and Personal communication[c]	X	X	X	X	
Boak and Mattson, report and personal communication[d]	X	X	X		
ExxonMobil	X	X	X	X	
URS[e]	X	X	X	X	X
AMEC[f]		X			
Shell EA, plan of operation and per-sonal communication[g]	X	X	X	X	X

Source: GAO analysis of selected studies.

[a] An "X" in the column indicates that we analyzed the water estimate in this study for this group of activities. We do not list quantitative water estimates for these groups of activities because some of these data are confidential.

[b] This study did not differentiate estimates for in-situ extraction from estimates for mining with a surface retort.

[c] This study was a confidential survey of multiple companies. Since we did not have access to the identities of the respondents or their individual answers, we could not exclude their estimates if they appear elsewhere in this table. The National Oil Shale Association published the survey results as total water needs being 1.7 barrels of water plus 0.6 to 1.6 barrels of water for upgrading liquids per barrel of oil produced. The upgrading estimate is for both in-situ and mining with a surface retort. Water for power generation and population growth is not included. We estimated water needs for power based on the average of the survey responses to power requirements. We estimated water needs for extraction and retorting and for reclamation based on the average of the survey responses. "Personal communication" indicates that we supplemented information in the study by contacting the author for more information.

[d] We included estimates for site water (water for extraction and retorting and reclamation) between the 25 percent and 75 percent cumulative probability levels. According to the author, about 90 percent of the site water is needed for reclamation. "Personal communication" indicates that we supplemented information in the study by contacting the author for more information.

[e] We could not separate water needs for upgrading liquids from extraction and retorting. The water needs for power are based on coal-fired plants.

[f] The purpose of this study is to update water requirements in the URS report. Preliminary data were presented by Benjamin Harding at the Promise and Perils of Oil Shale Symposium on February 5, 2010. Water needs for power are based on combined-cycle natural gas plants. The final study, which is expected to examine water needs for all groups of activities, will not be publicly available until October 2010.

[g] Shell cites a total of 3 barrels of water for each barrel of oil produced as appropriate for planning purposes. We estimated individual water needs for each of the five groups of activities by examining parameters discussed in Shell's EA and Plan of Operations in light of revised data provided verbally by Shell. Our estimates for individual groups of activities, based on Shell's revised data, add up to about 3 barrels of water. "Personal communication" indicates that we supplemented information in the study by contacting the author for more information.

Table 9. Studies GAO Examined That Contained Original Research on Water Requirements for Groups of Activities Representing the Complete Life Cycle for an Oil Shale Mine with a Surface Retort[a]

Study	Extraction and retorting and upgrading liquids[b]	Power	Reclamation	Population growth
BLM, PEIS				X
Dudley-Murphy, table 6, scenarios 2 and 6[c]				X
NOSA, report and personal communication[d]	X	X	X	
URS		X		X
OTA, Paraho-Direct process developed by WPA/DRI	X	X	X	X
OTA, Paraho-Direct process developed by McGee-Kunchal	X	X	X	X
OTA, Paraho-Indirect process developed by McGee-Kunchal	X	X		X
Oil Shale Exploration Company	X[e]	X[e]		

Source: GAO analysis of selected studies.

[a] An "X" in the column indicates that we analyzed the water estimate in this study for this group of activities. We do not list quantitative water estimates for these groups of activities because some of these data are confidential.

[b] We could not differentiate extraction and retorting from upgrading liquids in many of these studies.

[c] This study did not differentiate estimates for in-situ extraction from estimates for mining with a surface retort.

[d] This study was a confidential survey of multiple companies. Since we did not have access to the identities of the respondents or their individual answers, we could not exclude their estimates if they appear elsewhere in this table. The National Oil Shale Association published survey results as total water needs being 2 barrels of water plus 0.6 to 1.6 barrels of water for upgrading liquids per barrel of oil produced. The upgrading estimate is for both in-situ and mining with a surface retort. Water for population growth is not included. We estimated water needs for extraction and retorting and for reclamation based on the average of the survey responses. We estimated water needs for power based on the average of the survey responses to power needs. "Personal communication" indicates that we supplemented information in the study by contacting the author for more information.

[e] Although we reviewed these estimates, we excluded them from our analysis because we do not believe they are scalable to a commercial operation. They were moderately higher than the other estimates but not unreasonable, and they serve as a check on the upper limit for these two groups of activities.

We converted water requirements to barrels of water needed per barrel of oil produced. For those studies with water needs that met our criteria, we tabulated water needs for each group of activities for both in-situ production and mining with a surface retort. The results appear in tables 8 and 9. We estimated the total range of water needs for in-situ development by summing the minimum estimates for each group of activities and by summing the maximum estimates for the various groups of activities. We did the same for mining with a surface retort. We also calculated the average water needs for each group of activities.

To determine the extent to which water is likely to be available for commercial oil shale development and its source, we compared the total needs of an oil shale industry of various sizes to the amount of surface water and groundwater that the states of Colorado and Utah estimate to be physically and legally available, in light of future municipal and industrial demand. We selected the sizes of an oil shale industry based on input from industry and DOE. These are hypothetical sizes, and we do not imply that an oil shale industry will grow to these sizes. The smallest size we selected for an in-situ industry, 500,000 barrels of oil per day, is a likely size identified by an oil shale company based on experience with the development of the Canadian tar sands. The largest size of 2,500,000 barrels of oil per day is based on DOE projections. We based our smallest size of a mining industry, 25,000 barrels of oil per day, on one-half of the smallest scenario identified by URS in their work on water needs contracted by the state of Colorado. We based our largest size of a mining industry, 150,000 barrels of oil per day, on three projects each of 50,000 barrels of oil per day, which is a commonly cited size for a commercial oil shale mining operation. We reviewed and analyzed two detailed water studies commissioned by the state of Colorado to determine how much water is available in Colorado, where it was available, and to what extent demands will be placed on this water in the future.[39] We also reviewed a report prepared for the Colorado Water Conservation Board on future water availability in the Colorado River.[40] These studies were identified by water experts at various Colorado state water agencies as the most updated information on Colorado's water supply and demand. To determine the available water supply and the potential future demand in the Uintah Basin, we reviewed and analyzed data in documents prepared by the Utah Division of Water Resources.[41] We also examined data on water rights provided by the Utah Division of Water Rights and examined data collected by Western Resource Advocates on oil shale water rights in Colorado. In addition to reviewing these documents, we interviewed water experts at the Bureau of Reclamation, USGS, Utah Division of Water Rights, Utah Division of Water Resources, Utah Division of Water Quality, Colorado Division of Natural Resources, Colorado Division of Water Resources, Colorado River Water Conservation District, the Utah and Colorado State Demographers, and municipal officials in the oil shale resource area.

To identify federally funded research efforts to address the impacts of commercial oil shale development on water resources, we interviewed officials and reviewed information from offices or agencies within DOE and the Department of the Interior (Interior). Within DOE, these offices were the Office of Naval Petroleum and Oil Shale Reserves, the National Energy Technology Laboratory, and other DOE offices with jurisdiction over various national laboratories. Officials at these offices identified the Idaho National Laboratory and the Los Alamos National Laboratory as sponsoring or performing water-related oil shale research. In addition, they identified experts at Argonne National Laboratory who worked on the PEIS for BLM or who wrote reports on water and oil shale issues. Within Interior, we contacted officials with BLM and the USGS. We asked officials at all of the federal agencies and offices that were sponsoring federal research to provide details on research that was water-related and to provide costs for the water-related portions of these research projects. For some projects, based on the nature of the research, we counted the entire award as water-related. We identified 15 water-related oil shale research projects. A detailed description of these projects is in appendix II. To obtain additional details on the work performed under these research projects, we interviewed officials with all the sponsoring organizations and the performing organizations, including the Colorado School of Mines, University of Utah, Utah Geological Survey, Idaho National Laboratory, Los Alamos National Laboratory, Argonne National Laboratory, and the USGS.

Table 10. Agencies Contacted by GAO for Opinions on Research Needs

BLM
DOE Office of Naval Petroleum and Oil Shale Reserves (DOE NPOSR)
DOE National Energy Technology Laboratory (DOE NETL)
Bureau of Reclamation
USGS
Idaho National Laboratory
Los Alamos National Laboratory
Argonne National Laboratory
University of Utah
Colorado School of Mines
Colorado River Water Conservation District
Colorado Division of Water Resources
Utah Division of Water Quality
Colorado Geological Survey

Source: GAO.

To assess additional needs for research and to evaluate any gaps between research needs and the current research projects, we interviewed officials with 14 organizations and four experts that are authors of studies or reports we used in our analyses and that are recognized as having extensive knowledge of oil shale and water issues. The names of the 14 organizations appear in table 10. These discussions involved officials with all the federal offices either sponsoring or performing water-related oil shale research and state agencies involved in regulating water resources.

APPENDIX II. DESCRIPTIONS OF FEDERALLY FUNDED WATER-RELATED OIL SHALE RESEARCH

Research title	Sponsoring organization	Performing organization	Total federal cost	Federal cost related to water
Water Related Issues Affecting Conventional Oil & Gas Recovery and Oil Shale Development	DOE NETL	Utah Geological Survey	$688,223	$688,223[a]
GIS Water Resource Infrastructure for Oil Shale	DOE NETL	Colorado School of Mines	883,972	883,972[a]
Support for GIS Water Resource Infrastructure for Oil Shale	DOE NETL	Idaho National Laboratory	261,769	261,769[a]
Utah Center for Heavy Oil Research FY06[b]	DOE NETL	University of Utah	1,442,376	122,809c
Institute for Clean and Secure Energy FY08[b]	DOE NETL	University of Utah	873,340	154,937[c]
Institute for Clean and Secure Energy FY09[b]	DOE NETL	University of Utah	2,585,715	161,227[c]
Institute for Clean and Secure Energy FY10[b]	DOE NETL	University of Utah	3,044,800	160,160c
Carbon and Water Resources Impacts from Unconventional Fuel Development in the Western Energy Corridor	DOE NPOSR	Los Alamos National Lab	1,968,000	820,000[d]
Western Energy Corridor Initiative (support for Los Alamos)	DOE NPOSR	Idaho National Laboratory	500,000	100,000[e]
Dynamic Impact Model and Information System to Support Un- conventional Fuels Development	Idaho National Laboratory	Idaho National Laboratory	600,000f	250,000[e,f]
Generation and Expulsion of Hydrocarbons from Oil Shale	Idaho National Laboratory	Idaho National Laboratory	1,050,000f	90,000[e,f]
Near Field Impacts of In-Situ Oil Shale Develop-ment on Water Quality	Idaho National Laboratory	Idaho National Laboratory	612,500f	612,500[e,f]
Nuclear Pathways to Energy Security	Idaho National Laboratory	Idaho National Laboratory	75,000f	12,500[e,f]
Common Data Repository and Water Resource Assessment for the Piceance Basin, Western Colorado	BLM	USGS	110,000	110,000[a]
Water: Groundwater Monitoring in Piceance Basin and Yellow Creek Basin	BLM	USGS	410,000	410,000[a]
Total	15 projects		$15,105,695	$4,838,097

Source: DOE and Interior agencies and offices.

[a] Entire award is considered water-related due to the nature of the project.

[b] The University of Utah received four separate awards, each covering a broad array of oil shale research over multiple years. The awards included some water-related work. Examples of projects include (1) Meeting Data Needs to Perform a Water Impact Assessment for Oil Shale Development in the Uintah and Piceance Basins, (2) Effect of Oil Shale Processing on Water Compositions, and (3) New Approaches to Treat Produced Water and Perform Water Availability Impact Assessments for Oil Shale Development.

[c] DOE NETL provided this estimate of the water-related portion of the award.

[d] Los Alamos National Laboratory provided this estimate of the water-related portion of the award.

[e] Idaho National Laboratory provided this estimate of the water-related portion of the award.

[f] According to Idaho National Laboratory, some funding may be nonfederal, but it provided no details.

APPENDIX III. COMMENTS FROM THE DEPARTMENT OF THE INTERIOR

United States Department of the Interior
OFFICE OF THE SECRETARY
Washington, DC 20240

OCT 15 2010

Mr. Mark Gaffigan
Director, Natural Resources and Environment
Government Accountability Office
441 G Street, N.W.
Washington, D.C. 20548

Dear Mr. Gaffigan:

Thank you for the opportunity to review and comment on the Government Accountability Office (GAO) draft report entitled, *ENERGY-WATER NEXUS: A Better and Coordinated Understanding of Water Resources Could Help Mitigate the Impacts of Potential Oil Shale Development* (GAO-11-35). The Department of the Interior (DOI) generally concurs with GAO's findings and recommendations. The report focused on the research and coordination efforts of the Bureau of Land Management (BLM) and the United States Geological Survey (USGS) related to oil shale development.

The BLM and USGS continue to collaborate and collect baseline data in the Piceance Basin and to design and implement a groundwater and surface water monitoring network. Within the Piceance Basin, natural gas and oil development is occurring along with oil shale research and development activities. The USGS has compiled water quality data and continues to analyze groundwater and surface water to provide a baseline water quality assessment and identify data gaps and redundancies. This information can be used to understand current conditions, inform future monitoring in the basin, and as part of the evaluation performed under regulatory processes involved in the approval of large-scale commercial oil shale development.

The report recommends that the USGS and BLM "establish comprehensive baseline conditions for groundwater and surface water quality, including their chemistry, and quantity in the Piceance and Uintah Basins to aid in the future monitoring impacts from oil shale development in the Green River Formation." Implementation of this recommendation is underway. The USGS efforts for surface-water monitoring provide surface water quantity and quality monitoring at sites within these basins. A more comprehensive effort would include a work plan for gathering additional surface-water quantity data, groundwater monitoring, aquifer testing, and water-quality monitoring. In addition, the USGS Data Repository project, referred to in the report, will be used to better define data gaps in the baseline groundwater and surface water quality and quantity data currently existing for the Piceance and Uintah Basins.

The GAO next recommends that the BLM and USGS "model regional groundwater movement and the interaction between groundwater and surface water, in light of aquifer properties and the age of groundwater, so as to help in understanding the transport of possible contaminants derived from the development of oil shale." The BLM and USGS are working on shared needs for regional groundwater modeling. The Department agrees that data compilation and regional modeling should be performed prior to the approval of large-scale oil shale development. Modeling of the impact to regional groundwater and groundwater/surface water interaction requires accurate estimation of potential water use from the various oil shale development technologies as well as accurate baseline hydrologic information. The six new Research, Development and Demonstration (R, D&D) leases on Federal land in Colorado and Utah will allow industry to test various technologies to determine if production can occur in commercial quantities, and to develop an accurate determination of potential water use for each technology. Interior will then assess the economic and technological challenges involved with the R, D&D projects.

The third recommendation is for the BLM and USGS to "coordinate with the Department of Energy and state agencies with regulatory authority over water resources in implementing these recommendations, and to provide a mechanism for water-related research collaboration and sharing of results." Both bureaus are also working to improve such coordination. Currently, the bureaus are coordinating with state and local regulatory authorities in the many arenas of oil shale development and will build greater collaboration. As the results of the R, D&D leases become known, non-proprietary information will be shared within the Department and with Department of Energy.

If you have any questions, please contact LaVanna Stevenson-Harris, BLM Audit Liaison Officer, at 202-912-7088, or Rebecca Bageant, USGS Audit Liaison Officer, at 703-648-4328.

Sincerely,

Rhea S. Suh
Assistant Secretary
Policy, Management and Budget

APPENDIX IV. COMMENTS FROM THE DEPARTMENT OF ENERGY

Department of Energy
Washington, DC 20585

October 19, 2010

Mr. Mark E. Gaffigan
Director
Natural Resources and Environment Team
U.S. Government Accountability Office
441 G Street, NW, Mail 2T23A
Washington, DC 20548

Dear Mr. Gaffigan:

Thank you for the opportunity to review the Government Accountability Office (GAO) draft report entitled, *"Energy-water Nexus: A Better and Coordinated Understanding of Water Resources Could Help Mitigate the Impacts of Potential Oil Shale Development."* Enclosed pleased find the U.S. Department of Energy's comments on the draft report.

If you have any questions or comments please contact Mr. David F. Johnson, Deputy Assistant Director, Office of Petroleum Reserves, of my staff at (202) 586-4733.

Sincerely,

James J. Markowsky
Assistant Secretary
Office of Fossil Energy

Enclosure:
DOE Comments on Draft GAO Report

Department of Energy Comments On GAO Draft Report:
"Energy-Water Nexus: A Better and Coordinated Understanding of Water Resources Could Help Mitigate the Impacts of Oil Shale Development (GAO-11-35)"

The following comments are based on comments received from Idaho National Laboratory (INL), Los Alamos National Laboratory (LANL), and the Office of Petroleum Reserves (OPR):

1. **The Department of Energy (DOE) recognizes the need discussed in the Report for a more comprehensive and integrated cross-industry/government approach for determining water requirements, availability, and impacts associated with oil shale development (see for example Report *Highlights*).** Recognition of this need, as well as a potential solution, are discussed in the *Strategic Plan* described by Citation 25, a document that was prepared by representatives from industry, national laboratories, universities, and Federal, state, and local governments. Consideration should be given to including appropriate aspects of the *Strategic Plan* in the Report.

2. **Water requirements for oil shale in-situ technologies are overstated, and water available for an oil shale industry and all other uses is understated (see pages 17-21 and 25-36, and Tables 1, 3, 4, and 5, and Figure 4).**
   Water Requirements: The two primary in-situ processes are by Exxon Mobil Exploration Company (Exxon) and Shell Exploration and Production Company (Shell).

   - Exxon -- Exxon reports that life cycle water requirements are 1-2 barrels of water per barrel of oil produced, taking into account aggressive water conservation technologies (like utilizing produced water/process waste water, via water treating and recycling; using air cooled power generation; using natural gas instead of water for upgrading; and using natural gas instead of coal for power generation.)[1, 4, 7, 9]
   - Shell -- Shell representatives have stated publicly that life cycle water requirements have been studied, and that for planning purposes they are using 3 barrels per barrel, based on using aggressive water conservation technologies. They have also stated that ultimately they should be able to improve on this number.[2, 4, 20]
   - The Report dismisses water treating and recycling because the technologies have not been shown to be commercially viable (see page 16 of the Report). Challenges for oil shale are well within the bounds of commercially viable water treating/recycling technologies currently in operation in other industries in the U.S.; and, as such, the private sector is clearly planning to use them.[1, 3, 4, 7, 20]
   - The Report dismisses the use of produced water due to State-imposed legal limitations (see page 16 of the Report). Utilizing produced water for oil shale production is well within the bounds of how produced water is currently being used under these same limitations in the oil and gas and other industries in the U.S.; as such, the private sector is clearly planning to utilize produced water (within imposed legal limitations).[1, 3, 4, 6, 7, 20]
   - Thus, the water requirements range should be 1-3 barrels per barrel, not 0.8-12, as shown. The 12 barrel per barrel figure is based on a controversial report that was completed without sufficient industry input.[6] The primary complaint was that the URS report did not properly take into account aggressive utilization of water conservation technologies. The report has been reworked by AMEC Earth & Environmental Corporation (AMEC) with private industry input, and a draft of the AMEC report has

been publicly released showing a life cycle high number of 1.59 barrels per barrel. The AMEC report addressed comments received from National Oil Shale Association (NOSA), which incorporated comments from Exxon and Shell. [4, 5, 6, 7, 9, 20]

- On the basis of the foregoing, the Report should be revised to more fully describe water conservation technologies, with associated adjustments for in-situ water requirements.

<u>Water Availability</u> – The Report provides an analysis showing how using just the water available on the White River at Meeker would limit the size of an oil shale industry (see pages 32-36, and Tables 3, 4, and 5.). This implies that the size of an industry for the entire 3-State Green River Formation is limited by water available at Meeker. In actuality, water to support the entire industry would be drawn from *many* locations within the Upper Colorado River Basin. The Report should be revised to show how water available from the *entire* basin might limit a 3-State oil shale industry. Numerous data sources on Upper Colorado water flows are publicly available. See Citations 8, 9, 10, and 21.

3. **The Report gives the impression that all oil shale technologies are speculative, and that proving them to be commercially viable will be difficult, requiring a very long period of time, with uncertain outcomes (see pages 7 and 10). This is not an accurate represent-action of the state of the technologies, or of the expected timing of first commercial production. The Report should be revised to address these issues.**

- Advanced oil shale technologies have been under research for many years at a cost of billions of dollars. Much has been accomplished -- for some technologies; all, or almost all, technical questions impeding commercialization have been answered. [11, 12, 13, 14]
- The great majority of the components of the various oil shale technologies are not new; most have been in use in the oil and gas and other industries in the U.S. and elsewhere for many years -- e.g., mining, heating formations, fracturing formations, oil and gas production, environmental mitigation, upgrading, transporting, refining, etc. Much of the oil shale challenge is to integrate these components for oil shale in a manner that is economically and environmentally sustainable. This has been the primary focus of all of the research undertaken to date. Although this endeavor is very capital intensive, it is not technically difficult, and expectations have been generally optimistic, evidence of which is the billions of private research and development (R&D) dollars committed to date. The biggest obstacles to investment in the development of a viable oil shale industry in the U.S. have not been the state of the technology, but rather the regulatory uncertainty, and lack of access to resources on Federal lands in the western U.S. [11, 12, 13, 14, 22, 23, 25]

- Red Leaf Resources Inc (Red Leaf): The Report dismisses Red Leaf because it has not formally published detailed data on water needs (see page 50).
  - ➢ Red Leaf is a small start-up company that is funded with private capital. As such, it is not focused on publishing. In 2009, Red Leaf successfully completed a large field pilot/semi-works demonstration project. Based on this success Red Leaf has been funded to design and construct a 30,000 barrels per day commercial project that is due to begin operation within 2-3 years. Engineering is in progress. In order to permit its demonstration and commercial projects, detailed water studies were completed. Results have been presented at conferences and symposia. [15, 16, 18]

- Red Leaf's extraction process requires almost no water, and it is potentially applicable to any near-surface deposits, an enormous resource target. [15, 16, 18]
- Oil Shale Exploration Company's (OSEC) surface retort technology has been in commercial operation in Brazil since 1992, and is proven there. Production capacity is about 4,600 barrels per day. [10, 17]
  - All that remains for OSEC to do is adapt the technology as needed to make it economic to operate in the U.S. regulatory environment.
  - OSEC completed a feasibility study in 2009 which found no need for a demonstration project and estimated operating costs to be $29-35 per barrel. [10, 17]
  - OSEC is designing a 12,500 barrel per day commercial plant that is funded and expects to begin operation in 2016, expanding to 50,000 barrels / day by 2027. [10, 17]
  - The OSEC technology is potentially applicable to any near-surface or mineable deposits, an enormous resource target. [10, 17]
- There are several other technologies that are also promising, especially the technology by Enshale Inc.[19]
- The Report should be revised to more fully address Red Leaf, OSEC, and other technologies.
- The Report should also more fully reflect the extensive history of oil shale research in the U.S., including ongoing public and private efforts. [11, 24, 25]

FOOTNOTES/CITATIONS

1. Thomas, Michele M., et al. (Exxon Mobil Exploration Company): *"Responsible Development of Oil Shale"* Presented to 29th Oil Shale Symposium, October 19-21, 2009.
2. US General Accountability Office: *"Energy-Water Nexus: A Better and Coordinated Understanding of Water Resources Could Help Mitigate the Impacts of Oil Shale Development (GAO-11-35)* pg. 52, footnote "g".
3. Shell Exploration and Production Company *"Plan of Operations – Oil Shale Test Project"* Prepared for US Bureau of Land Management, February 15, 2006.
4. Harding, Ben (AMEC Earth & Environmental Corporation) *"Energy Development Water Needs Assessment and Water Supply Alternative Analysis"* presentation to Yampa River Basin Roundtable Meeting, July 21, 2010.
5. National Oil Shale Association *"Position Paper: Studies Misrepresent Future Oil Shale Impacts"* April 2009. www.oilshaleassoc.org
6. United Research Services (URS) Corp. *"Energy Development Water Needs Assessment (Phase I Report)"*, prepared for Colorado, Yampa, and White River Basin Roundtables Energy Subcommittee. September 2008.
7. Harding, Benjamin (AMEC): *"Oil Shale Direct Water Use Estimates – DRAFT"* Memorandum to Joint Energy Water Needs Subcommittee, April 13, 2010.
http://www.crwcd.org/media/uploads/Water_and_Energy_Oil_Shale_Direct_Water_Use_Final_DRAFT_20100413.pdf
8. Wood, Thomas, et al. *"Environmental Aspects of CO2 and Water Management"* Presentation to 27th Oil Shale Symposium, Colorado School of Mines. October 15-17, 2007. Slides 7-9.
9. Harding, Benjamin (AMEC) *"Energy Development Water Needs Assessment and Water Supply Alternative Analysis: The Promise and Peril of Oil Shale"* February 5, 2010. Slides 9-13.
10. Aho, Gary (Oil Shale Exploration Company) *"White River Oil Shale Project 2009 Update"* Presentation to 29th Oil Shale Symposium, October 19-21, 2009.
11. *"Secure Fuels from Domestic Resources: The Continuing Evolution of America's Oil Shale and Tar Sands Industries"* prepared by INTEK, Inc. for US Department of Energy, September 2010.
12. *Strategic Significance of America's Oil Shale Resource; Volume I Assessment of Strategic Issues and Volume II Oil Shale Resources, Technology and Economics.* Prepared for the U.S. Department of Energy Office of Naval Petroleum and Oil Shale Reserves, March 2004.
13. US Department of Energy *"Joint USA –Estonia Oil Shale Research and Utilization program: Phase I, Phase II, and Phase III Reports and Executive Summary"* Compilation of Reports, 2009.
14. US Congress, Office of Technology Assessment "An Assessment of Oil Shale Technology" June 1980.
15. Patten, James, et al (Red Leaf Resources Inc) *"The EcoShale Process: Field Pilot Construction, Operation, and Results"* Presentation to 29th Oil Shale Symposium, October 19-21, 2009.
16. *"Secure Fuels from Domestic Resources: The Continuing Evolution of America's Oil Shale and Tar Sands Industries"* prepared by INTEK, Inc for US Department of Energy September 2010. – Red Leaf company profile.
17. *"Secure Fuels from Domestic Resources: The Continuing Evolution of America's Oil Shale and Tar Sands Industries"* prepared by INTEK, Inc for US Department of Energy September 2010. – OSEC company profile.
18. Dammer, Anton (Red Leaf Resources Inc) *"The EcoShale In-Capsule Retorting Technology"* Presentation to International Oil Shale Symposium, Tallinn, Estonia June 2009.
19. *"Secure Fuels from Domestic Resources: The Continuing Evolution of America's Oil Shale and Tar Sands Industries"* prepared by INTEK, Inc. for US Department of Energy, September 2010. – Enshale Inc profile.
20. *"Energy Water Use Scenarios"* Memorandum to Joint Energy Water Needs Subcommittee, June 29, 2010.
21. US Bureau of Land Management: *"Final Programmatic Environmental Impact Statement (PEIS)"*, September 2008. http://www.ostseis.anl.gov/eis/guide/index.cfm
22. Task Force on Strategic Unconventional Fuels, *Initial Report of Findings and Recommendations: Development of America's Strategic Unconventional Fuels Resources,* September 2006.
23. Task Force on Strategic Unconventional Fuels, *Development of America's Strategic Unconventional Fuels; Volumes I- III,* September 2007.
24. *"Oil Shale Research in the United States: Profiles of Oil Shale Research and Development Activates in Universities, National Laboratories, and Public Agencies"* prepared by INTEK, Inc. for US Department of Energy, September 2010.
25. Ad Hoc Unconventional Fuels Working Group: "Strategic Plan: Unconventional Fuels Development in the Western Energy Corridor" November 2008.

End Notes

[1] The Rand Corporation, a nonprofit research organization, estimates that between 30 and 60 percent of the oil shale in the Green River Formation can be recovered. At the midpoint of this estimate, almost half of the 3 trillion barrels of oil would be recoverable.

[2] GAO, *Energy-Water Nexus: Improvements to Federal Water Use Data Would Increase Understanding of Trends in Power Plant Water Use*, GAO-10-23 (Washington, D.C.: Oct. 16, 2009) and *Energy-Water Nexus: Many Uncertainties Remain about National and Regional Effects of Increased Biofuel Production on Water Resources*, GAO-10-116 (Washington, D.C.: Nov. 30, 2009).

[3] Physically available, according to the state of Colorado, is the actual or observed amount of water flowing in a stream. This amount can vary from year to year, based on the amount of precipitation and snow pack. Legally available, according to the state of Colorado, is the portion of physically available flow that could be developed without injury to other water rights or compacts.

[4] An aquifer is an underground layer of rock or unconsolidated sand, gravel, or silt that will yield groundwater to a well or spring.

[5] Reclamation is an attempt to mitigate the adverse impacts of heating the subsurface zone, such as repeated rinsing with water to remove any residual hydrocarbons that were not economically extracted.

[6] For a detailed discussion of the literature we reviewed and the experts to whom we spoke, see appendix I.

[7] In directional drilling, the company starts drilling a borehole on the disturbed ground surface and angles the well so that the bottom of the hole occurs below the undisturbed surface.

[8] Porosity is the amount of space within an aquifer that can be filled with groundwater. Permeability is the ability of a material, such as an aquifer or rock formation, to transmit liquids like water.

[9] In Shell's original process, a ring of bore holes is drilled around the zone to be isolated. Liquid ammonia is circulated down the boreholes, which freezes the groundwater in the immediate vicinity, creating a ring of ice around the isolated zone.

[10] One barrel contains 42 gallons.

[11] The state of Colorado has promulgated extensive regulations regarding the nature of water produced from oil and gas operations. According to Colorado state officials, the transport and use of this water offsite to oil shale operations may be restricted.

[12] For a complete list of the studies we reviewed and a detailed description of our methodology, see appendix I.

[13] Shell and the URS Corporation—a contractor to the state of Colorado—conducted the two studies that examine water needs for all five groups of activities comprising the life cycle of in-situ oil shale development. For planning purposes, Shell cites 3 barrels of water needed per barrel of oil. URS estimates that 5.2 barrels of water would be needed per barrel of oil.

[14] We calculated that AMEC's estimated power requirements exceeded by over seven times the coal-fired generating capacity of northwest Colorado, which consists of this nearest plant and one other smaller plant.

[15] The National Oil Shale Association provided this estimate for upgrading shale oil. This range also contains data for upgrading oil from surface retorts, which we could not segregate. Conversations with an oil shale industry representative suggest that water estimates for upgrading oil derived from in-situ operations may lie toward the bottom of this range.

[16] More information on studies we examined appears in appendix I. Experts consider some of the data within the OTA study to still be relevant because certain surface retort technologies are similar to those being tested today.

[17] The two studies that examined water needs for all five groups of activities that comprise the life cycle of oil shale development by mining and surface retorting are included in the OTA report. Both studies involve the Paraho–Direct Process. These estimates are 2.3 and 2.8 barrels of water per barrel of oil.

[18] We multiplied these power requirements by the amounts of water needed to generate power as the amounts appear in GAO, *Energy-Water Nexus: Improvements to Federal Water Use Data Would Increase Understanding of Trends in Power Plant Water Use*, GAO-10-23 (Washington, D.C.: Oct. 16, 2009), p. 20.

[19] Western Resource Advocates, *Water on the Rocks: Oil Shale Water Rights in Colorado* (Boulder, Colo., 2009).

[20] A conditional water right is a water right that has not yet been put to beneficial use. Its date of application establishes its priority among other water rights.

[21] An acre-foot is the amount of water that would fill an area of one acre to a depth of one foot. An acre-foot contains 325,851 gallons, or 7,758 barrels, and is roughly equal to the amount of water that a family of four will use in a year.

[22] An absolute water right is a water right that has been put to beneficial use.

[23] Shell also experienced considerable formal opposition to its proposal from 25 groups, some of which was for environmental reasons.

[24] Year-to-year flows on rivers can vary significantly with annual precipitation. However, officials with the state of Colorado said that they are comfortable using average annual flows.

[25] This estimate represents all of the groundwater and surface water that can be used in the Uintah Basin, but does not take into account any current withdrawals from streams and rivers.

[26] Many research projects involve water and nonwater issues. For projects that include nonwater-related segments, we obtained estimates of the amount of the project spent on water related tasks.

[27] Most projects run for 2 to 3 years. Some have been completed, while others are still ongoing.

[28] A general description of the process DOE uses to select research proposals can be found in GAO, *Research and Development: DOE Could Enhance the Project Selection Process for Government Oil and Natural Gas Research,* GAO-09-186 (Washington, D.C.: Dec. 29, 2008).

[29] In January 2010, BOR initiated the *Colorado River Basin Water Supply and Demand Study* at a federal cost of $1 million. Although not directed at oil shale, this 2-year study's objective is to define and resolve current and future imbalances between the supply and demand for water within the Colorado River Basin over the next 50 years.

[30] NEPA requires all federal agencies to consider the environmental impacts of their actions and decisions. It requires an analysis and a detailed statement of the environmental impact of any proposed major federal action which significantly affects the environment.

[31] 40 C.F.R. §1505.2 (c).

[32] *The NEPA Task Force Report to the Council on Environmental Quality: Modernizing NEPA Implementation* (September 2003).

[33] USGS, Colorado Water Science Center, *Regional Framework for Water-Resources Monitoring Related to Energy Exploration and Development* (Sept. 30, 2007).

[34] GAO, *Results-Oriented Government: Practices that Can Help Enhance and Sustain Collaboration among Federal Agencies,* GAO-06-15 (Washington, D.C.: Oct 21, 2005).

[35] GAO, *Watershed Management: Better Coordination of Data Collection Efforts Needed to Support Key Decisions,* GAO-04-382 (Washington, D.C.: June 7, 2004).

[36] National Research Council of the National Academies, *Confronting the Nation's Water Problems: The Role of Research* (2004).

[37] Subcommittee on Water Availability and Quality, National Science and Technology Council, *A Strategy for Federal Science and Technology to Support Water Availability and Quality in the United States* (Washington, D.C., September 2007).

[38] This research was funded by ExxonMobil, Shell, and Total Exploration and Production.

[39] CDM, *Statewide Water Supply Initiative,* a report contracted by the Colorado Water Conservation Board, November 2004; and CDM *Colorado's Water Supply Future: State of Colorado 2050 Municipal and Industrial Water Use Projections,* a report contracted by the Colorado Water Conservation Board, June 2009.

[40] AECOM *Colorado River Water Availability Study (Draft Report),* a report contracted by the Colorado Water Conservation Board, March 2010.

[41] Utah Division of Water Resources, *Utah's Water Resources: Planning for the Future* (May 2001); *Municipal and Industrial Water Supply and Uses in the Uintah Basin (Data Collected for Calendar-Year 2005)* (December 2007); and *Utah State Water Plan, Uintah Basin* (December 1999).

CHAPTER SOURCES

The following chapters have been previously published:

Chapter 1 – This is an edited reformatted and augmented version of a United States Government Accountability Office publication, report GAO-09-862T, dated July 9, 2009.

Chapter 2 - This is an edited reformatted and augmented version of a United States Government Accountability Office publication, report GAO-10-23, dated October 2009.

Chapter 3 - This is an edited reformatted and augmented version of a United States Government Accountability Office publication, report GAO-10-116, dated November 2009.

Chapter 4 - This is an edited reformatted and augmented version of a United States Government Accountability Office publication, report GAO-11-35, dated October 2010.

INDEX

#

21st century, 9

A

aboveground tank systems, 75
access, 16, 23, 27, 32, 53, 106, 108, 138, 139
accounting, 4, 45, 55, 128
accounting standards, 4
acid, 68
adhesives, 76
adverse effects, 2, 8, 12, 19, 25, 32, 51, 78, 129
age, 129, 131, 132
agencies, 3, 5, 7, 9, 12, 14, 15, 22, 24, 39, 43, 45, 46, 48, 50, 53, 58, 61, 62, 66, 72, 86, 89, 97, 99, 100, 102, 104, 129, 130, 131, 132, 133, 136, 140, 141, 142, 149
agricultural producers, 89
agricultural residues, 4, 5, 72, 95
agriculture, 5, 66, 69, 80, 81, 82, 86, 89, 91, 96
air emissions, 28
air pollutants, 9, 106
air quality, 62, 96, 102, 106
air temperature, 29, 60
Alaska, 95
alcohols, 9
algae, vii, 1, 4, 6, 64, 65, 71, 73, 76, 83, 84, 85, 86, 95, 96
alternative water sources, 2, 7, 8, 11, 12, 14, 15, 17, 24, 25, 27, 32, 33, 34, 37, 38, 43, 46, 47, 50, 53, 58, 64, 78, 82
ambient air, 29, 60
ambient air temperature, 29
ammonia, 32, 148
amphibians, 91

Annual Energy Outlook, 3, 24, 95
aquaculture, 95
aquatic life, 95, 106
aquifers, 6, 64, 77, 78, 86, 95, 100, 107, 109, 110, 118, 120, 122, 129, 131
assessment, 14, 45, 53, 97
atmosphere, 13, 19, 84, 90, 96
audit, 4, 16, 49, 102
authorities, 15
authority, 3, 38, 45, 54, 62, 78, 104, 132, 136

B

background information, 89
bacteria, 32, 71, 95
barriers, 63, 64, 66, 78, 81, 82, 88, 96
benefits, 8, 12, 14, 15, 24, 25, 27, 28, 30, 33, 42, 47, 69, 72, 84, 87, 90, 91
benzene, 75, 96
bias, 133
biochemical processes, 67
biodiesel, 4, 65, 66, 74, 75, 85, 89, 96
biofuel, vii, 2, 3, 4, 5, 6, 63, 64, 65, 66, 67, 69, 71, 74, 75, 76, 78, 79, 81, 82, 83, 85, 86, 87, 88, 89, 95, 96
biofuels, vii, 1, 2, 3, 4, 5, 6, 59, 63, 64, 65, 67, 68, 69, 72, 73, 74, 76, 77, 78, 83, 85, 87, 88, 89, 95, 96
biomass, 64, 67, 73, 76, 87, 95
biorefineries, 4, 6, 64, 67, 69, 74, 75, 78, 82, 85, 86, 87
biotechnology, 81
Biotechnology Industry Organization, 66, 89
birds, 91
blends, 6, 67, 76, 85, 96
boilers, 74, 75
boreholes, 108, 148

bounds, 133
Brazil, 105
breeding, 96
Bureau of Land Management, 100, 101, 132, 135, 136, 137
burn, 111

C

California Environmental Quality Act, 53
canals, 50
carbon, 24, 29, 68, 72, 77, 95, 96, 114, 128
carbon dioxide, 29, 68, 77, 114
case studies, 15, 48
case study, 15, 47, 89
cash, 81
cash flow, 81
cattle, 68
CEC, 12, 33, 53, 54, 55
cellulose, 67
cellulosic biofuel, 73, 95
Cellulosic feedstocks, 4
certificate, 36, 38, 39, 56, 61
certification, 48, 50, 51, 53, 56
challenges, 2, 7, 8, 16, 25, 26, 32, 82, 101, 105, 106, 131, 134
chemical, 32, 67, 69, 73, 75, 86, 96
chemical characteristics, 86
chemicals, 32, 75, 91, 108
China, 105
cities, 13, 128
clarity, 71
cleaning, 26
climate, vii, 24, 25, 33, 63, 65, 100, 101, 118, 122, 125, 128
climate change, vii, 24, 63, 65, 122, 125, 128
climates, 29, 60, 122
coal, 3, 4, 8, 9, 13, 15, 25, 26, 29, 30, 31, 34, 47, 49, 55, 59, 60, 61, 106, 111, 113, 117, 133, 138, 139, 148
coatings, 91
coding, 89
collaboration, 88, 130, 132
combustion, 44
commerce, 14, 95
commercial, viii, 4, 5, 9, 25, 64, 72, 73, 74, 83, 84, 96, 99, 102, 104, 105, 110, 124, 132, 133, 134, 135, 136, 140
commercial-scale facilities, 4
communication, 131, 139, 140

communities, 13, 14, 74, 102, 105, 121
compatibility, 6, 76, 85
competition, 134
compilation, 129
compliance, 12, 24, 32, 53
compounds, 96
computer, 54, 124
conceptual model, 129
configuration, 28, 47
conflict, 35
Congress, 14, 45, 85, 131
consent, 14
conservation, 5, 35, 36, 45, 56, 63, 64, 65, 78, 79, 80, 81, 85, 88, 90
constituents, 32, 96
construction, 56, 62, 100, 105, 106, 107
consumers, 29
consumption, 4, 5, 7, 11, 12, 13, 14, 20, 22, 24, 26, 27, 42, 43, 44, 45, 46, 56, 58, 60, 62, 70, 74, 82, 84, 115, 122, 137
consumption rates, 26
contaminant, 74, 75
contamination, 75
content analysis, 89
contingency, 56
conversion of feedstocks, vii, 1, 68, 69
cooking, 67
cooling, vii, 1, 2, 3, 4, 5, 7, 8, 11, 12, 13, 14, 15, 16, 17, 18, 19, 20, 21, 22, 24, 25, 26, 27, 28, 29, 30, 31, 32, 33, 34, 35, 36, 37, 38, 39, 40, 42, 43, 44, 45, 46, 47, 48, 50, 51, 52, 53, 54, 55, 56, 57, 58, 60, 61, 62, 64, 65, 68, 74, 75, 82, 85, 95, 96, 110, 111, 114, 138
cooling process, 5, 7
cooling technologies, vii, 2, 4, 7, 8, 11, 12, 15, 18, 22, 24, 25, 27, 28, 30, 31, 33, 34, 37, 38, 39, 42, 43, 46, 47, 48, 50, 53, 55, 57, 61, 111
coordination, 46, 97, 100, 130, 131, 133
copper, 32
corn stover, 4, 5, 72, 84
corrosion, 32
cost, 2, 4, 7, 8, 16, 27, 29, 31, 32, 46, 51, 60, 61, 64, 68, 72, 79, 119, 129, 142, 149
covering, 66, 89, 142
crises, 103
crop, 2, 5, 64, 67, 70, 71, 72, 77, 79, 80, 81, 82, 83, 90, 91, 95, 96
crop production, 72
crop residue, 67, 72, 79, 80, 90

crops, 2, 4, 5, 69, 71, 72, 73, 76, 79, 81, 82, 83, 84, 85, 90, 96, 122
CRP, 64, 72, 77, 84, 87
crude oil, 95, 101, 103, 105, 134
cultivation, vii, 1, 2, 3, 5, 6, 63, 64, 66, 68, 69, 70, 71, 72, 73, 74, 76, 77, 78, 79, 80, 81, 82, 83, 84, 85, 86, 87, 88, 89, 95, 96
customers, 24
cycles, 37, 61, 62

D

damages, 61
data collection, 2, 8, 11, 15, 43, 46, 58, 85
data gathering, 8
database, 43, 46, 54, 129
decomposition, 96
degradation, 77, 109
denial, 61
Department of Agriculture, 3, 65, 66, 89, 92
Department of Commerce, 48, 62, 66, 89
Department of Energy, viii, 2, 3, 8, 12, 13, 47, 48, 60, 65, 93, 96, 99, 100, 101, 132, 136, 144
Department of the Interior, viii, 14, 59, 89, 94, 99, 100, 101, 141, 143
Department of Transportation, 96
Departments of Agriculture, 63
deposits, vii, 74, 99, 101, 102, 104, 123
depth, 66, 86, 88, 89, 96, 120, 148
diminishing returns, 61
discharges, 5, 18, 27, 32, 37, 51, 59, 75, 84, 104, 110
diseases, 90
dissolved oxygen, 109
distillation, 65
distribution, 2, 3, 6, 12, 44, 46, 50, 68, 75, 81, 85, 86, 91, 135
diversity, 66, 73, 89
DOI, 63, 86, 87
draft, 60, 63, 86, 96, 132, 134
drainage, 71, 91, 110
drinking water, 74, 75
drought, 5, 56, 65, 76, 81, 83, 101, 105
drying, 75

E

ecology, 66, 89
economic downturn, 119
economic growth, 13, 121
economics, 14
education, 81
effluent, 8, 12, 17, 25, 32, 33, 37, 46, 49, 50, 51, 61, 82, 88
electricity, vii, 1, 2, 3, 4, 7, 11, 12, 13, 14, 15, 16, 22, 24, 25, 26, 27, 28, 29, 30, 31, 33, 36, 39, 45, 46, 47, 48, 49, 52, 55, 58, 62, 106, 109, 110, 111, 113, 117, 133, 138
elk, 106
emergency, 103
emission, 24, 90
employees, 32
endangered species, 39, 52
energy, vii, 1, 2, 3, 4, 6, 7, 9, 12, 14, 15, 22, 28, 29, 30, 31, 34, 44, 47, 51, 53, 60, 61, 63, 67, 81, 82, 85, 91, 96, 102, 103, 104, 105, 129
Energy Independence and Security Act, 9, 65
Energy Information Administration's (EIA), 3, 95
Energy Policy Act of 2005, 67, 104, 129, 133
energy supply, 14, 22
enforcement, 128
engineering, 4, 26, 31, 47, 66, 81, 86, 89, 96
enrollment, 81
environment, 3, 6, 7, 15, 22, 25, 27, 32, 44, 74, 75, 95, 101, 149
environmental conditions, 129
environmental control, 26
environmental effects, 39, 50, 87, 91
environmental factors, 16
environmental impact, viii, 4, 51, 53, 81, 85, 87, 99, 100, 101, 102, 104, 128, 149
environmental issues, 16, 38
environmental organizations, 89
Environmental Protection Act, 39
Environmental Protection Agency (EPA), 3, 6, 13, 28, 29, 48, 54, 56, 59, 60, 63, 65, 66, 87, 89, 95, 96, 100, 104, 108, 136
environmental quality, 66, 89
environmental regulations, 14, 57
enzymes, 67, 68
equipment, 2, 4, 5, 6, 8, 12, 20, 25, 28, 29, 32, 37, 74, 76, 81, 82, 87, 96, 111
erosion, 72, 73, 79, 84, 90, 91, 110
Estonia, 105
ethanol, 2, 4, 5, 6, 9, 64, 65, 66, 67, 68, 70, 71, 72, 73, 74, 75, 76, 78, 82, 83, 85, 87, 89, 95, 96
Ethanol, 4, 6, 9, 67, 69, 70, 75, 77, 95, 96
ethanol-blended fuels, 75, 76, 96
evaporation, 18, 19, 26, 29, 32, 60, 61, 62, 70, 74, 90, 91, 96

evapotranspiration, 70, 83, 84
evidence, 4, 16, 49, 66, 90, 102
executive orders, 103
expertise, 82, 127
exposure, 71, 90
extraction, 3, 6, 32, 86, 100, 107, 108, 110, 111, 113, 115, 116, 123, 136, 137, 138, 139

F

families, 105, 112, 115
farm land, 73
farmers, 3, 5, 71, 73, 76, 80, 81, 82, 84, 85, 122
farmland, 85
farms, 13
federal agency, 40, 42, 66, 89, 102, 126, 129
federal government, vii, viii, 3, 14, 45, 50, 63, 65, 67, 88, 99, 100, 101, 103, 104, 126, 131
federal lands, viii, 14, 99, 101, 103, 134
federal law, 14, 54, 104
federal water data, vii, 11, 15, 42, 45, 47, 48
feedstock, vii, 2, 4, 5, 6, 63, 64, 65, 66, 67, 68, 69, 72, 73, 74, 75, 76, 77, 78, 79, 83, 84, 86, 87, 88, 89
fermentation, 4, 65, 68, 83
fertility, 95
fertilizers, 71, 73, 91, 95
fiber, 68
field trials, 84
filtration, 96
financial, 101, 103
financial support, 101, 103
fish, 24, 35, 71, 91, 108, 110, 122
fishing, 105
flexibility, 7, 12, 25, 27, 75
flooding, 91
flour, 68
fluctuations, 29
football, 31
forest management, 73
formation, 106, 148
fractures, 110
freshwater, vii, 2, 3, 5, 6, 7, 8, 9, 11, 12, 13, 14, 15, 16, 17, 24, 25, 27, 28, 32, 33, 37, 42, 43, 44, 45, 47, 61, 64, 65, 70, 73, 75, 76, 78, 82, 84, 85, 95, 96
fuel consumption, 28, 30
funding, 14, 34, 44, 45, 81, 82, 127, 128, 142
funds, 41, 126, 127, 131

G

GAO, vii, viii, 1, 2, 3, 11, 12, 16, 18, 19, 20, 21, 36, 41, 44, 54, 57, 58, 59, 63, 64, 66, 87, 89, 90, 91, 95, 96, 97, 99, 100, 106, 112, 113, 114, 116, 117, 121, 124, 126, 127, 135, 137, 138, 139, 141, 148, 149
Georgia, 15, 34, 35, 36, 39, 40, 41, 47, 49, 55, 56, 57, 66, 76, 78, 89
glycerin, 75, 96
Glycerin, 75, 96
governments, 3, 105, 131
grades, 116
grading, 108
grants, 78
grasses, 64, 67, 71, 72, 73, 76, 79, 80, 81, 84
greenhouse, 24
groundwater, viii, 6, 8, 12, 17, 22, 23, 25, 34, 35, 36, 37, 38, 39, 40, 41, 46, 49, 50, 51, 52, 54, 56, 57, 62, 68, 70, 71, 74, 75, 78, 86, 96, 99, 100, 102, 107, 108, 109, 110, 113, 118, 119, 120, 125, 126, 127, 128, 129, 130, 131, 132, 134, 135, 140, 148, 149
growth, 4, 32, 56, 65, 71, 74, 95, 105, 112, 113, 116, 138, 139
guidance, 55
Gulf of Mexico, 71, 77, 95

H

habitat, 18, 27, 39, 52, 72, 87, 91, 106
hardwoods, 9
harvesting, 72, 84
Hawaii, 95
hazards, 96
haze, 106
health, 13, 51, 85
heavy metals, 84
hemicellulose, 67
highways, 105
history, 119, 134
host, 103
House, 13, 47, 65, 101, 131
House of Representatives, 13, 65, 101, 131
housing, 105, 115, 137
human, 45
human behavior, 45
humidity, 29
hunting, 105

hybrid, 18, 21, 25, 26, 28, 30, 46, 54, 57, 60, 61
hydrocarbons, 83, 96, 108, 110, 111, 113, 148
hydroelectric power, 15, 47
hydrogen, 96

I

Idaho National Laboratory (INL), 127
identification, 7, 127
illusion, 133
imbalances, 149
Impact Assessment, 137, 142
impairments, 69, 71, 73
imports, 52
improvements, 12, 58, 82
impurities, 8, 25
income, 81
incompatibility, 2
independence, 4, 9
individuals, 89, 104
industries, 3, 13, 15, 58, 69, 102, 105, 123
industry, viii, 1, 7, 9, 12, 14, 15, 16, 18, 19, 20, 21, 37, 39, 41, 42, 43, 44, 45, 46, 47, 48, 49, 66, 76, 78, 81, 82, 87, 89, 96, 99, 100, 102, 104, 106, 107, 114, 115, 117, 118, 121, 122, 123, 124, 125, 127, 131, 132, 133, 134, 136, 140, 148
infancy, 131
infrastructure, 2, 6, 83, 85, 96, 105
injure, 61
injury, iv, 148
insects, 90
institutions, 72, 88, 104
invertebrates, 71, 108, 109
investment, 101, 131
investments, 24
Iowa, 66, 70, 71, 82, 85, 86, 89
irrigation, 2, 5, 8, 22, 25, 46, 62, 65, 66, 70, 71, 73, 76, 81, 83, 86, 89, 91, 95, 96, 119, 121
issues, vii, 2, 3, 6, 12, 15, 32, 35, 36, 43, 47, 48, 49, 51, 53, 61, 62, 63, 65, 75, 126, 128, 129, 130, 141, 149

J

jurisdiction, 141

L

lakes, 35, 65, 71, 79
landscape, 73, 77, 105
landscapes, 71
laws, 1, 12, 14, 15, 22, 24, 27, 34, 35, 48, 50, 53, 78
leaching, 71, 90, 91, 108
lead, 5, 6, 12, 28, 32, 50, 53, 57, 71, 73, 81, 86
leakage, 76
leaks, 6, 76, 83, 85, 108
legislation, 14, 85
life cycle, 65, 68, 84, 87, 100, 102, 110, 111, 112, 113, 115, 116, 117, 133, 136, 137, 148
lifetime, 25
light, 4, 130, 132, 139, 140
liquid fuels, 87, 95
liquid phase, 121
liquids, 95, 113, 116, 138, 139, 148
livestock, 95
local authorities, 133
local government, 3
logging, 76
lumber mills, 4, 73

M

magnitude, 33, 60, 61, 72, 100, 107, 136
majority, 13, 44, 52, 65, 69, 126, 130
management, 9, 45, 55, 62, 77, 90, 91, 131
Maryland, 49
materials, 65, 109
matter, iv, 15, 48
metals, 32, 108, 110
methanol, 96
methodology, 15, 58, 60, 61, 66, 88, 89, 148
Mexico, 104
miscanthus, 4
mission, 22, 44
Mississippi River, 71, 77, 95
Missouri, 70
models, 41, 54, 84, 122, 124, 130
moisture, 50, 68, 72, 79, 81, 95
moisture content, 81
molecules, 67
mortality, 108, 110
municipal solid waste, 73

N

National Aeronautics and Space Administration, 66, 89
National Environmental Policy Act (NEPA), 128

national parks, 106
National Pollutant Discharge Elimination System, 51, 54, 56, 64, 75
National Research Council, 45, 62, 73, 83, 84, 95, 96, 131, 149
natural enemies, 73
natural gas, 4, 13, 15, 26, 27, 29, 30, 31, 46, 47, 49, 51, 52, 61, 112, 113, 117, 120, 125, 139
natural resources, 105
negative effects, 72
neutral, 22
New York, iv
next generation, 2, 4, 5, 6, 64, 65, 69, 74, 76, 81, 83, 85, 86, 87
nitrogen, 71, 73, 80, 83, 90, 91, 106
North America, 49, 137
nutrient, 2, 5, 64, 71, 72, 73, 77, 79, 80, 81, 83, 84, 90, 91, 95, 96
nutrients, 71, 72, 73, 79, 80, 81, 84, 90, 91, 96

O

officials, vii, viii, 3, 5, 7, 8, 11, 12, 14, 22, 36, 37, 38, 39, 40, 41, 42, 43, 44, 45, 48, 52, 53, 56, 57, 58, 60, 61, 62, 63, 64, 66, 70, 71, 72, 74, 75, 76, 77, 78, 79, 81, 82, 83, 84, 85, 86, 89, 96, 99, 100, 102, 118, 120, 121, 122, 125, 126, 128, 129, 130, 136, 140, 141, 148
oil, vii, 1, 3, 4, 6, 8, 25, 63, 65, 69, 85, 99, 100, 101, 102, 103, 104, 105, 106, 107, 108, 109, 110, 111, 112, 113, 115, 116, 117, 118, 119, 120, 121, 122, 123, 124, 125, 126, 127, 128, 129, 130, 131, 132, 133, 134, 135, 136, 137, 138, 139, 140, 141, 142, 148, 149
Oil Shale Exploration Company, 101, 118, 135, 137, 139
operating costs, 61
operations, 14, 76, 100, 101, 106, 107, 108, 109, 110, 111, 112, 113, 115, 117, 118, 120, 121, 123, 124, 127, 133, 148
opportunities, 130
organic compounds, 110, 116
organic matter, 90
organism, 71
osmosis, 75, 96
overlap, 57, 123
oversight, 34, 35, 66, 89
ownership, 23, 52, 85, 118, 119
oxygen, 71, 95, 96, 109, 110

P

Pacific, 49, 66, 76, 89
pathogens, 84
penalties, 12, 28, 29, 30, 31, 34, 60, 61
perennial energy crops, 4
permeability, 109, 110, 113
permission, iv
permit, 12, 35, 36, 38, 39, 51, 52, 55, 56, 62, 74, 75, 78, 96
personal communication, 139
pesticide, 2, 5, 64, 71, 73, 79, 81, 90, 91, 96
pests, 73
petroleum, 4, 65, 66, 67, 75, 96, 103
Petroleum, 9, 70, 95, 96, 102, 103, 126, 127, 128, 133, 141
phosphates, 32
phosphorous, 95
phosphorus, 71
pilot-scale basis, 4
plant type, 31
plants, vii, 2, 3, 4, 5, 6, 7, 8, 9, 11, 12, 13, 14, 15, 16, 17, 18, 19, 21, 22, 24, 25, 26, 27, 29, 30, 31, 32, 34, 37, 38, 39, 42, 43, 44, 45, 46, 47, 48, 50, 51, 52, 53, 54, 55, 56, 57, 60, 61, 62, 64, 68, 70, 71, 74, 76, 82, 89, 91, 96, 108, 111, 113, 117, 125, 133, 138, 139
policy, 22, 37, 38, 50, 53, 54, 55, 61, 118
policy making, 22
policymakers, 42, 43, 44, 45, 58
pollutants, 27, 29, 32, 75, 84, 91, 104, 109
pollution, 5, 20, 53, 62, 90
ponds, 4, 21, 22, 26, 82, 109
population, 3, 15, 47, 49, 52, 55, 65, 101, 111, 112, 115, 117, 121, 125, 136, 137, 138, 139
population growth, 3, 55, 65, 111, 115, 121, 136, 137, 138, 139
porosity, 109, 110, 113
potential benefits, 7
power generation, 35, 39, 55, 111, 115, 117, 138
power lines, 108
power plants, vii, 1, 2, 3, 4, 7, 8, 9, 11, 12, 13, 14, 15, 16, 17, 21, 22, 24, 25, 27, 28, 31, 32, 34, 35, 36, 37, 38, 39, 40, 41, 42, 43, 44, 45, 46, 47, 50, 51, 52, 53, 54, 55, 56, 58, 59, 61, 62, 82, 111, 117
precipitation, 16, 24, 50, 55, 68, 70, 71, 108, 109, 122, 148
predators, 84
preparation, iv, 91, 108
President, 103, 131

Index

prevention, 90
primary data, 40
probability, 139
producers, 4, 89
profit, 81
programmatic environmental impact, 101, 104
project, 33, 101, 103, 107, 127, 128, 142, 149
property rights, 23
protection, 80, 118
public concern, 8, 33
public health, 53
public interest, 35
pumps, 28, 30, 31, 60

Q

quality control, 54, 58
quality of life, 13
quality standards, 14, 32, 106

R

rainfall, 76, 78
recommendations, iv, 1, 11, 45, 46, 63, 99, 129, 132, 133
recovery, 80, 95, 96, 104
recreation, 13, 95
recycling, 74, 111, 113, 128, 133, 134, 137
regulations, 8, 24, 27, 32, 53, 104, 128, 148
regulatory agencies, 48, 102, 127, 130, 131
regulatory oversight, 34, 37
regulatory requirements, 8, 33
reliability, 61, 88
remediation, 96
renewable energy, 52
renewable fuel, 2, 4, 65, 67, 85
Renewable Fuel Standard, 65
Renewable Fuels Association, 66, 89
rent, 85
reproduction, 109
requirements, 3, 4, 16, 28, 34, 36, 37, 39, 40, 51, 53, 54, 56, 61, 66, 68, 73, 78, 81, 83, 88, 95, 96, 113, 117, 122, 125, 128, 138, 139, 140, 148
research institutions, 15, 48
researchers, 4, 5, 6, 66, 74, 89, 126, 127, 128, 129, 130
reserves, vii, 99, 101, 103
residuals, 73
residues, 4, 5, 72, 79, 90, 95

resource management, 66, 89
resources, vii, 1, 3, 4, 9, 13, 15, 22, 44, 45, 64, 66, 69, 83, 87, 89, 100, 102, 103, 104, 106, 107, 108, 109, 126, 128, 129, 130, 131, 134, 136, 138
response, vii, 55, 63, 65, 87, 102, 133, 134
restoration, 91
restrictions, 35, 36, 61, 111, 122
retail, 24, 62
revenue, 29, 31
RFS, 65, 67, 95
rights, 23, 35, 40, 41, 52, 62, 104, 118, 119, 120, 121, 123, 125, 140
risk, 6, 40, 83, 96
risks, 75, 87, 96
root, 84, 91
root system, 84
royalty, 105
runoff, 5, 64, 69, 71, 72, 73, 77, 79, 80, 81, 83, 90, 91, 108, 122
rural areas, 112

S

safety, 26, 31, 53
saline water, 17, 43, 113
salts, 8, 25, 75, 96, 108, 110
savings, 8, 30, 33, 37, 61
scaling, 32
school, 105
science, 22, 45, 66, 89, 131
scope, 15, 66, 87, 134
security, 9
sediment, 5, 73, 77, 84, 90, 91, 108, 109
sedimentation, 73, 95
sediments, 71
seed, 90, 91
selenium, 108, 110
sellers, 4
Senate, 131
services, iv
sewage, 2, 8, 17, 25, 32, 37, 61, 82, 110
short supply, 8
shortage, 23
skilled workers, 112
society, 15
soil erosion, 72, 73, 79, 90, 95
soil type, 96
solid waste, 32
sorghum, 4
South Dakota, 70, 78

soybeans, 4, 65, 67, 71
species, 14, 73, 106, 122
specifications, 30, 61
stakeholders, 43, 66, 89, 126
starch, 65, 67, 68
state, vii, 1, 3, 11, 12, 14, 15, 22, 24, 27, 34, 35, 36, 37, 38, 39, 40, 41, 42, 45, 47, 48, 49, 50, 51, 52, 53, 54, 55, 56, 58, 61, 62, 63, 66, 71, 72, 75, 76, 77, 78, 82, 84, 86, 88, 89, 95, 97, 99, 100, 102, 104, 107, 112, 114, 118, 120, 121, 122, 123, 124, 125, 126, 129, 130, 131, 132, 133, 134, 140, 141, 148
state laws, 24, 36, 54
state oversight, 12
state regulators, vii, 11, 15, 35, 36, 37, 40, 41, 47, 48, 78
states, vii, 5, 11, 12, 14, 15, 23, 24, 34, 35, 37, 38, 39, 41, 42, 47, 49, 52, 53, 55, 61, 63, 66, 67, 70, 72, 74, 76, 78, 82, 86, 89, 95, 104, 120, 121, 122, 132, 134, 140
statistics, 60, 61
statutes, 36
storage, 2, 3, 6, 50, 65, 75, 85, 89, 96, 104, 118, 119, 122, 123, 128
stress, vii, 63, 105
substitutes, 96
substitution, 44
sulfur, 106
sulfur dioxide, 106
supplier, 37
surface area, 106, 108
sustainability, 87
switchgrass, vii, 1, 4, 72, 76

T

tanks, 6, 75, 76
tar, 140
target, 84, 110
technical comments, 11, 63, 87
technical support, 81
techniques, 91, 96
technological advancement, 2, 6
technologies, 2, 3, 4, 7, 8, 9, 11, 12, 14, 15, 18, 22, 24, 25, 27, 28, 30, 31, 33, 34, 38, 39, 42, 43, 45, 46, 47, 48, 50, 53, 57, 61, 64, 67, 81, 82, 85, 100, 104, 105, 107, 110, 111, 131, 132, 134, 136, 138, 148

technology, 2, 4, 5, 6, 7, 8, 13, 14, 20, 22, 25, 26, 30, 31, 33, 39, 43, 44, 46, 53, 55, 56, 60, 74, 82, 84, 85, 101, 104, 105, 111, 114, 131, 132
temperature, 26, 29, 30, 60, 61, 109, 110
Tennessee Valley Authority, 48
testing, 91, 132, 134
thermoelectric power plants, vii, 1, 3, 5, 7, 9, 11, 12, 13, 14, 15, 16, 22, 24, 37, 38, 42, 43, 44, 45, 46, 47, 50, 53, 55, 59, 111
thinning, 73
time periods, 58
total costs, 31, 61
total energy, 29
tourism, 105
toxic substances, 108, 110
toxicity, 75
trade, 9, 15, 25, 30, 32, 33, 34, 47, 81
trade-off, 9, 15, 25, 30, 32, 33, 34, 47
training, 82
transmission, 16, 27, 33, 106
transparency, 131
transpiration, 96
transport, vii, 1, 3, 27, 30, 75, 76, 87, 91, 108, 119, 122, 129, 130, 132, 148
transportation, 4, 65, 66, 67, 95
treatment, 2, 8, 17, 25, 26, 27, 32, 37, 61, 68, 75, 76, 82, 91, 110, 128, 133

U

U.S. Army Corps of Engineers, 48, 56
U.S. Geological Survey, 4, 12, 13, 14, 17, 40, 48, 51, 54, 56, 65, 66, 89, 95, 101, 132, 135, 136
uncertain outcomes, 134
underground storage tanks (UST), 75
United, 2, 4, 7, 11, 13, 14, 16, 17, 18, 22, 25, 27, 55, 62, 63, 64, 65, 66, 69, 70, 77, 88, 95, 96, 97, 99, 102, 105, 128, 131, 134, 135, 137, 149
United States, 2, 4, 7, 11, 13, 14, 16, 17, 18, 22, 25, 27, 55, 62, 63, 64, 65, 66, 69, 70, 77, 88, 95, 96, 97, 99, 102, 105, 128, 131, 134, 135, 137, 149
universe, 7, 21
universities, 4, 41, 49, 102, 104, 129
updating, 60
uranium, 4, 13
urban, 71, 73, 101
urban areas, 73
USDA, 3, 6, 63, 65, 66, 70, 79, 81, 82, 83, 84, 86, 89, 96

V

variables, 45
variations, 24, 52, 61, 62
varieties, 5, 64, 73, 81, 83
vegetation, 79, 91, 106, 108, 109
vegetative cover, 72
vehicles, 67
visualization, 130

W

Washington, 59, 60, 62, 95, 96, 148, 149
waste, 2, 61, 73, 75, 76, 107, 108, 109, 110, 111, 112, 113
waste heat, 112
waste water, 2, 61, 107, 110, 113
wastewater, 6, 8, 13, 25, 52, 53, 75, 76, 82, 91, 96
water quality, 2, 3, 5, 6, 9, 32, 51, 53, 54, 56, 61, 62, 64, 66, 69, 71, 72, 73, 74, 75, 77, 78, 79, 83, 84, 88, 89, 102, 108, 109, 110, 126, 127, 128, 129, 132
water quality standards, 84
water resources, vii, viii, 3, 6, 14, 15, 24, 36, 41, 42, 44, 45, 63, 64, 65, 66, 69, 72, 73, 78, 79, 83, 85, 86, 87, 88, 97, 99, 100, 102, 104, 106, 107, 108, 126, 128, 129, 130, 131, 132, 133, 136, 140, 141

water rights, 23, 35, 40, 50, 52, 53, 62, 104, 118, 119, 120, 121, 122, 123, 124, 125, 126, 127, 140, 148
water shortages, 13, 36, 52, 86
water supplies, 2, 3, 4, 5, 7, 8, 12, 27, 37, 40, 53, 55, 65, 70, 73, 75, 76, 78, 86, 100, 125
watershed, 35, 58
waterways, 5, 35, 91, 122, 123
wealth, 105
well-being, 131
wells, 35, 36, 51, 71, 78, 108, 109, 110, 111, 113, 128, 129, 137
wetlands, 51, 79
wilderness, 106
wildlife, 13, 27, 32, 35, 72, 87, 91, 102, 105, 106
Wisconsin, 70
withdrawal, 14, 17, 18, 20, 22, 35, 40, 41, 51, 55, 56, 57, 62, 78, 86, 106, 120, 128
wood, 9, 73
wood products, 73
wood waste, 73
workers, 105, 112, 115, 137

Y

yield, 4, 54, 68, 70, 81, 148